THE SCHOLARLY PUBLISHING OFFICE
THE UNIVERSITY OF MICHIGAN
UNIVERSITY LIBRARY

COLLECTIONS OF MINERALS, ROCKS, AND FOSSILS.

METALLURGIC MINERALOGY.—COLLECTIONS of ORES of the Principal METALS, with a Catalogue of Names and Localities, *without Cabinets.*

12. One hundred Specimens, three square inches, 4*l.* 4*s.*
13. One hundred Specimens, four square inches, 5*l.* 5*s.*

COLLECTIONS of SPECIMENS of SIMPLE MINERALS, arranged in Systematic Order. The specimens are well selected, fresh, clean, characteristic, and accurately named. *Without Cabinets.*

14. Two hundred Specimens, size four square inches, 5*l.* 5*s.*
15. Three hundred Specimens, size four square inches, 8*l.* 8*s.*
16. Four hundred Specimens, size six square inches, 14*l.*
17. Five hundred Specimens, size six square inches, 20*l.*

CRYSTALLOGRAPHY.—A Series of ONE HUNDRED AND TWENTY MODELS OF CRYSTALS.

These Models represent the most important Natural Crystals, both of simple and complicated forms. Their size is from 2 to 4 inches in diameter. The material of which they are formed is cream-coloured biscuit Porcelain, which presents the following advantages :—The models are much *stronger* than those made of paper. Their edges are sufficiently sharp, and their planes sufficiently even, to permit very good *approximate measurements* to be taken by means of the common goniometer. They can be *written upon*, either with a black-lead pencil, the marks from which can be effaced by India-rubber, or with common ink, which is easily removable by muriatic acid. Consequently, the names of the Crystals, or their symbols, or the angles across their edges, or the names of the Minerals they represent, can be written upon them and removed at pleasure. These properties are not possessed in the same degree by models made of glass or wood. When soiled, they can be *cleaned* by soap and water. Finally, they are *cheaper* than models made of any other material.

Price of Selections of the Models of Crystals :—

18. First Selection, 40 Models, 21*s.*
19. Second Selection, 40 Models, 21*s.*
20. Third Selection, 40 Models, 21*s.*
21. The Collection of 120 Models, 2*l.* 12*s.* 6*d.*

GRIFFIN'S GEOLOGICAL CABINETS.

The Cabinets of Rocks are arranged precisely in the same manner as the Cabinets of Minerals already described, and they present to the Geologist the same advantages that those Cabinets do to the Mineralogist. The only difference between them is that the specimens of Rocks are larger than the Minerals at the same cost.

22. Price of 150 Specimens, in a Cabinet, 3*l.* 13*s.* 6*d.*
23. Price of 100 Specimens, in a Cabinet, 2*l.* 2*s.*

COLLECTIONS OF FOSSILS, arranged in the order in which they occur in the Strata composing the crust of the earth, with a Catalogue.

These collections contain Fossils belonging to all formations, and from many localities, and in such variety as could only be obtained by the assiduous researches of many years, attended by very great expense. The utmost pains have been taken by a competent Naturalist to determine and name the specimens with accuracy.

Prices of the Collections, with Catalogues, *without Cabinets:—*

24. One hundred Specimens, 3*l.* 3*s.*
25. One hundred and fifty Specimens, 5*l.* 5*s.*
26. Two hundred Specimens, 8*l.*
27. Three hundred Specimens, 15*l.*
28. Five hundred Specimens, 25*l.*

GEOLOGICAL COLLECTIONS, containing Specimens of ROCKS, with their CHARACTERISTIC FOSSILS. *Without Cabinets.*

These collections are adapted either for private study or for the use of lecturers, and have been prepared with great attention to their scientific accuracy. The Rocks are disposed in the order of their natural position relative to one another; and the Fossils, being those most characteristic of the different Rocks, and in number about one-third of the entire collection, are placed just after the Rocks to which they belong. The collections are accompanied with catalogues. The size of the cut specimens of Rocks is 9 square inches, namely, 3½ inches by 2½ inches, very neatly cut, fresh, and clean.

29. Collection of Five hundred Specimens, 20*l.*
30. Collection of Three hundred Specimens, 12*l.*
31. A Choice General Collection, of medium size, six square inches, containing 500 Minerals, 500 Rocks, 500 Fossils, and 120 Models of Crystals, in all 1,620 Specimens, 60*l.*

ELEMENTS

OF

MINERALOGY AND GEOLOGY.

BY

FRIEDRICH SCHOEDLER, PH. D.

PROFESSOR OF THE NATURAL SCIENCES AT WORMS, AND FORMERLY ASSISTANT
IN THE CHEMICAL LABORATORY OF GIESSEN.

EDITED FROM THE FIFTH GERMAN EDITION

BY

HENRY MEDLOCK, F.C.S.

SENIOR ASSISTANT IN THE ROYAL COLLEGE OF CHEMISTRY.

ILLUSTRATED BY NUMEROUS ENGRAVINGS ON WOOD.

LONDON:

PUBLISHED BY JOHN JOSEPH GRIFFIN AND CO.,

53 BAKER STREET, PORTMAN SQUARE,

AND RICHARD GRIFFIN AND CO., GLASGOW.

1851.

ADVERTISEMENT.

The following Outline of the Sciences of Mineralogy and Geology forms a Section of an Elementary Introduction to the Natural and Physical Sciences, published under the title of the Book of Nature.

This Section is published apart, for the accommodation of persons actively engaged in the study of Rocks and Minerals, who may be desirous of having the portions which relate to those subjects presented in the most convenient form.

It will be understood that the occasional references to *Physics, Chemistry, &c.*, are to the other Sections of the Book of Nature.

CONTENTS.

I.—MINERALOGY.

II.—GEOLOGY.

MINERALOGY AND GEOLOGY.

1. MINERALOGY is the science which treats of MINERALS, or those constituents of the earth which are of similar composition through their entire mass.

All parts of the same Mineral are alike. In none of them do we observe those peculiar structures which we term *Organs*, and which, in Plants and Animals, fulfil functions that are indispensable to the existence of the individuals to which they belong. Minerals are consequently also called *Inorganic* substances. It is immaterial whether the mineral subjected to examination is large or small; in either case it presents to us the same structure. A small specimen of sandstone gives us as good an idea of its properties as would be given by a large block or a mountain of the same material. A rock crystal a line long is as perfect as one that measures a foot.

2. In the section Chemistry (§ 3 and § 9) we have already seen that the entire mass of the earth is composed only of a few more than 60 simple substances or elements. By virtue of the chemical affinity possessed by these bodies they are found to exist in the greatest varieties of combination, and only seldom occur as simple substances. Proceeding from this view, Mineralogy might be termed the science of the chemical compounds occurring in Nature, and this is, to a certain extent, really the case. We have already, in studying the science of Chemistry, become acquainted with a number of such natural compounds, and have been referred to this section for a description of some others.

Chemical affinity, however, is not the only power by which the elements are influenced in the great laboratory of Nature. Its action is accompanied by that of a number of forces and influences, producing a series of mineralogical formations which cannnot be considered simply from a chemical point of view, nor explained by chemical affinity alone.

3. Minerals may, therefore, be classed in two principal groups, which are readily distinguishable from each other. Those of the first class possess all the properties of perfect chemical compounds, are definite in their chemical composition, and equally definite in their crystalline form. These bodies are the true or *simple* Minerals, and the science that treats of them is MINERALOGY.

The other series of minerals has an essentially different character. The objects composing it are either evident mixtures of simple minerals, or, if they agree with simple minerals in chemical composition, they differ by the absence of definite crystalline form. Hence they do not occur as well-defined individuals, but in amorphous masses. These substances are called *Mixed Minerals*, *Stones*, or *Rocks*. The study of their individual properties, of their relations to each other and to the mass of the earth, and the investigation of their origin and mode of aggregation, constitute the subject of that division of this science which is termed GEOLOGY.

I.—MINERALOGY.

4. The first condition required of Mineralogy is that it shall furnish us with means to recognise Minerals and determine the classes to which they belong. Minerals may be distinguished and classified by certain characteristics. These are, principally, 1, their *Form*; 2, their *Physical*, and 3, their *Chemical*, properties. With the aid of these characteristics we can make an attempt to describe minerals.

1. FORM OF MINERALS.—CRYSTALLOGRAPHY.

5. We have already remarked in Physics (§ 20), and in Chemistry (§ 29), that the smallest particles of chemical compounds attract each other and arrange themselves in certain directions, producing regularly-formed bodies, which are called *crystals*. These forms exhibit *faces* or *planes*; *edges* or lines of contact of two planes, and *points* or *angles* formed by the meeting of three or more planes. There is no form of crystal exhibiting less than four planes, six edges and four angles, and most crystals are composed of a larger number.

As every mineral, with few exceptions, always crystallises in one definite primary form, the observation of their forms is naturally a very important and certain mode of discriminating minerals. The forms of crystals are, however, exceedingly numerous. On examining a collection of minerals, hundreds of different forms are presented to the eye; yet these varieties may be traced back to a few fundamental forms, from which they are all derived. Of these fundamental forms there are six, which, with the various secondary or derived forms, constitute six families or systems of crystallisation, the study of which forms a particular branch of science termed *Crystallography*. It is not possible for us to enter into the details of this science, but we will endeavour to make ourselves acquainted with the primary forms, with the most important secondary forms, and with the manner in which a secondary crystal is described and traced back to its primary form.

Primary Forms of Crystals.

6. The *regular octohedron* (fig. 1). This crystal is limited by eight equal equilateral triangles, and has twelve edges and six angles. An imaginary line, extending from one angle to the opposite one, represents what is called the *axis* of the crystal. The octohedron has therefore three such axes, which are all equal and intersect each other at right angles. The whole form of the crystal is dependent upon this relation of the axes. If we construct a cross with three knitting-needles of equal length, so that they cross each other at right angles, their ends will represent the angles of a regular octohedron.

1.

Irregular octohedrons, which are likewise primary forms, may easily be represented by such crosses The axes are either of unequal length, or they do not intersect each other at right angles, or they may exhibit both of these deviations.

In examining and describing a crystal, it is always placed in such a position that one of the axes is situated vertically in front of the observer; this is called the *principal axis*, while the others are termed *secondary* axes. When the axes of a crystal are equal, any one may be adopted as the principal axis, but when they are unequal the longest is generally chosen as the principal axis.

7. The *secondary* forms of the octohedron, as of all other crystals, are produced by the real or imaginary removal of certain portions of the primary form in a regular manner. We will give a few examples of this, which may be rendered more lucid by the use of prepared models or by cutting the requisite forms out of a piece of potato or turnip. On removing the angles

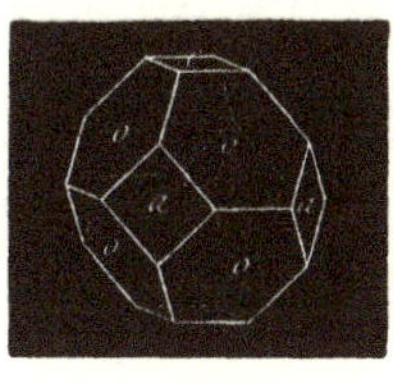

2.

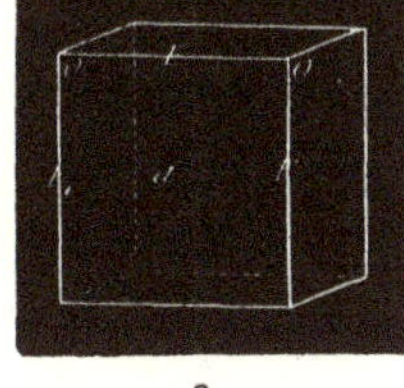

3.

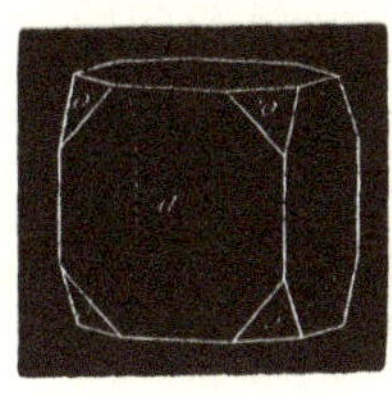

4.

from the octohedron (fig. 2) by parallel sections, a cube will at last be obtained. The *cube*, or six-faced solid (fig. 3), has six square planes of equal magnitude, eight angles, and twelve edges. On removing the angles of this crystal, as in fig. 4, the regular octohedron is obtained. The crystal represented by

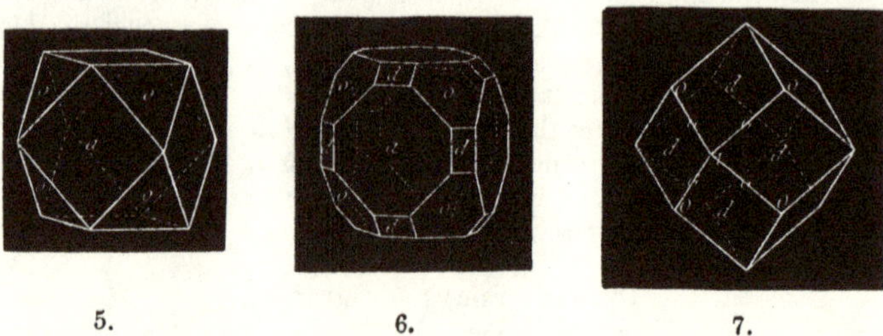

5. 6. 7.

fig. 5 is intermediate between those shown by figures 2 and 4, or between figures 1 and 3. It is evident that these forms bear certain definite relations to each other, and therefore they are said to belong to the same system of crystallisation, which has been called the *regular* system.

A number of secondary forms may be obtained by truncations that remove the angles and edges of the primary forms. Thus, fig. 6 is a cube deprived of its edges and angles. On the removal of the edges of the cube in a regular manner the *rhombic dodecahedron* is obtained (fig. 7), of which the twelve equal planes are rhombs.

Another series of secondary forms—the *hemihedral* forms—is produced by removing, not the whole of the angles or edges of a primary form, but only

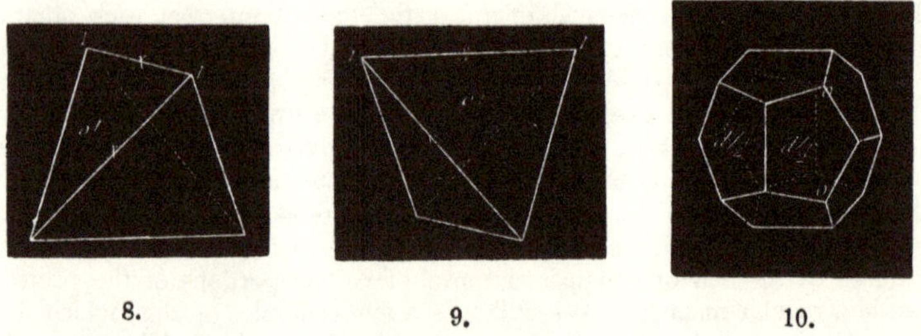

8. 9. 10.

the alternate angles or edges which are situated opposite to each other. The three-sided pyramid, or *tetrahedron* (figs. 8 and 9), is thus obtained from the octohedron. The *pentagonal dodecahedron* (fig. 10) is a secondary form obtained in a somewhat similar manner.

8. The *second* primary form is the *quadratic octohedron* (fig. 11): this form has three axes intersecting each other at right angles, two of which are equal, while the third is longer or shorter than the others. The horizontal section of this octohedron is a square. The vertical section is a rhomb. By truncation of its lateral edges the prism (fig. 12) is obtained. The latter may have its edges and angles replaced by various modifications.

The *third* primary form is the *rhombic-octohedron* (fig. 13), the axes of which cross each other at right angles, but are all unequal. The three sections of this octohedron are rhombs. This is one of the crystalline forms of sulphur. The rhombic *prism* is obtained by the truncation of the lateral edges of the octohedron, namely, those that are marked G in the figure.

The *fourth* primary form is an octohedron with three axes of unequal length, two of which intersect each other at an oblique angle, but are placed

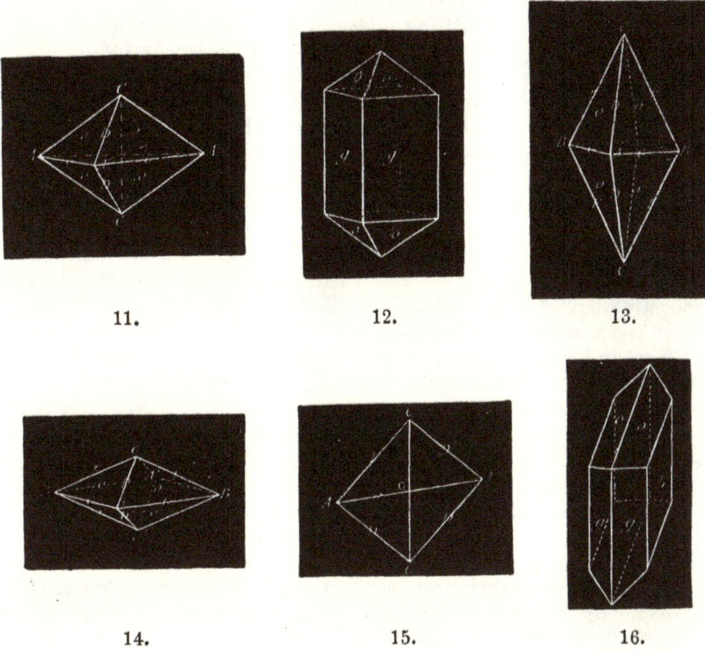

11. 12. 13.

14. 15. 16.

at right angles to the third axis, as shown in figs. 14, 15. This octohedron generally occurs in its secondary forms, particularly in oblique rhombic prisms, as in gypsum (fig. 16).

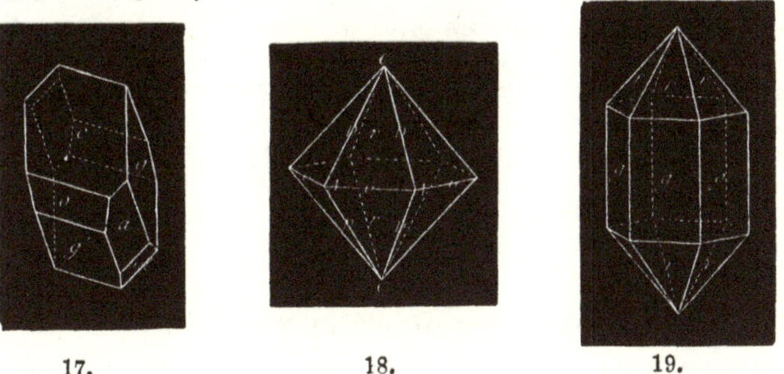

17. 18. 19.

The *fifth* primary form is an octohedron of which all the axes are unequal, and all intersect each other at oblique angles. This octohedron only occurs in its secondary forms, such as is exhibited by fig. 17, which is the crystalline form of axinite.

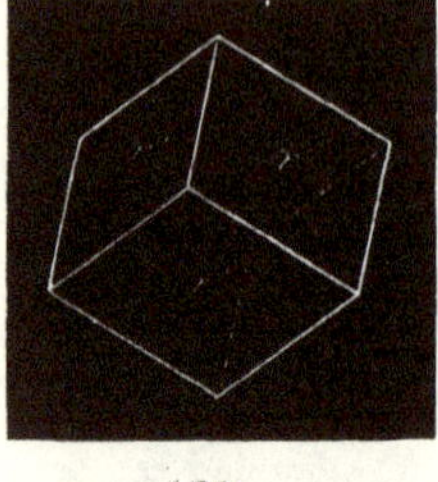

20.

The *sixth* primary form is the *hexagonal dodecahedron* (fig. 18), a double six-sided pyramid. This system, which is termed the *hexagonal system*, is the only one which has four axes: three of these are equal, and intersect each other at acute but equal angles. The fourth or principal axis is unequal to the others and intersects them at a right angle. Among the secondary forms of this system may be mentioned the beautiful hexagonal prism (fig. 19), and the rhombohedron (fig. 20), which is enclosed by six equal rhombs.

9. The same mineral frequently occurs in very different crystalline forms, which, however, will always be found to belong to one of these six systems, that is to say, they may always be traced back in some way or other to primary forms included in these systems. The determination of the form of the crystal is frequently attended with considerable difficulty. This arises in part from the similarity which many forms have to each other, so that they can frequently be distinguished only by the most accurate measurements of the edges and angles of the crystals. Sometimes the difficulty is caused by the imperfect nature of the crystals that are to be examined; for crystals so regularly and distinctly formed as they are usually exhibited in diagrams, are of rare occurrence. In attempting to account for the irregularities of natural crystals, it may generally be assumed that there were obstacles present at the instant of their formation which did not permit the crystal to be equally and perfectly developed on all sides. Thus, it is frequently observed that only one-half, or only an edge, an angle, or a plane of the crystal is formed, the remainder of the form being either entirely wanting or merged in an extraneous adhering mass (see the figure at page 216). The most regular forms frequently appear to be perfectly irregular, the necessary conditions for the formation of the crystal having been favourable only on one side of the mass. The difficulties attending the examination of crystals may, however, be overcome by practice. Models of crystals, which may easily be procured, are of great assistance in the study of crystallography.

A variety of familiar names are given to irregular crystalline formations, such as tables, plates, needles, &c.

A mineral appears as a *crystalline* mass or aggregate when it consists of small crystals irregularly and closely arranged together. Thus, for instance, calcareous spar is distinctly crystallised, and marble only crystalline, carbonate of lime. When a mineral exhibits no crystalline arrangement of its particles, it is said to be massive or amorphous. Cavities in many masses of rocks are sometimes filled with groups of crystals. These are called *geodes* or *drusic cavities* (comp. § 81).

2. PHYSICAL CHARACTERS OF MINERALS.

10. As minerals cannot always be distinguished by their form, other characteristic properties are called into aid, such as the cohesion, the specific gravity,

and the colour of minerals, and their relations to light, electricity, and magnetism. These are termed the *physical characters* of minerals.

Cohesion.

11. There are only two liquid minerals; the greater number is solid, and with these special regard is paid to their cleavage, fracture, and hardness.

A mineral is cleavable when it is possessed of crystalline structure. In such a case its particles are arranged in a certain manner, so as to exhibit less cohesive power in one direction than in another, in the same manner as wood is more easily cleavable in the direction of its fibre than across the grain. There are, of course, various degrees of *cleavability*; thus, mica is cleavable into the thinnest laminæ. Cleavage surfaces are more or less plane.

Fractures, or fractured surfaces, are produced by the forcible disintegration of minerals that are not cleavable, or of cleavable minerals in any other direction than that of their cleavage. The appearance of the fracture is very characteristic of many minerals; it is either *even* or *uneven*, or it may be *conchoidal*, as for example, in the case of flint. It is also *splintery*, or *jagged*, and very frequently *earthy*, as in chalk and many other minerals.

12. In the description of a mineral, particular attention is always paid to its *hardness*. Many minerals are sufficiently hard to resist the best files, while others are so soft as to admit of being scratched with the finger nail. There are between these two extremes various degrees of hardness, which cannot be so easily described. Other means have, therefore, been resorted to for the determination of the degree of hardness of different minerals with tolerable accuracy. Of two minerals, that one of course is the hardest which will scratch the other without being scratched itself. A scale of hardness has been constructed by Mohs, consisting of ten well-known minerals, so arranged that each one will scratch that which precedes it, and may be itself scratched by all those which follow it in the scale. Thus, ten degrees of hardness are obtained between the softest mineral, which is talc, and the hardest, namely diamond: these degrees are represented by the corresponding numbers; they are as follows:—

Degree of hardness 1 = Talc.
2 = Gypsum.
3 = Calcareous spar.
4 = Fluorspar.
5 = Apatite.

Degree of hardness 6 = Felspar.
7 = Quartz.
8 = Topaz.
9 = Corundum.
10 = Diamond.

If, therefore, the degree of hardness of a certain mineral is 7, we know it to be equal to that of quartz. It may easily be remembered that a high degree of hardness is represented by a high number, and *vice versâ*. The best way of trying the hardness of a mineral is to rub it on a hard fine-toothed file in comparison with the minerals that constitute the scale, pieces of which are reserved for that use. The scratching of minerals by one another is liable to several fallacies.

Specific Gravity of Minerals.

13. The density or specific gravity of a body, as was shown in Physics (§ 53), is the weight of a certain volume, compared with that of an equal

volume of water. Thus, the density of lead is = 11, since 1 cubic inch of this metal weighs 11 times as much as 1 cubic inch of water. We have already spoken of the value of a knowledge of specific gravities; the fact of substances always possessing the same density under uniform circumstances, furnishes an important means of recognizing them, particularly in the case of minerals. Hence the determination of their specific gravities has been made and repeated with the greatest care, generally at a temperature of 15°·5 C. (60° F.). From what has been said in Chemistry, it may in general be assumed that minerals of high specific gravity, contain the heavy metals.

Relation of Minerals to Light.

14. As minerals are a very extensive and diversified class of bodies, they exhibit an exceedingly varied behaviour under the influence of the rays of light. Many minerals permit these rays of light to pass through them, refracting or deflecting them in the passage. Others reflect the light in a peculiar manner. These properties give rise to the transparency, the refracting power, the lustre, and the colour of minerals.

Transparency is either perfect or imperfect. We meet with the former chiefly in highly-developed crystals; and when this character appears in conjunction with the absence of colour, it is termed simply *transparent*. Imperfect transparency is indicated by the terms *semi-transparency* and *translucency;* pellucid at the edges. The term *opaque* means that no rays of light are transmitted by the mineral.

Refracting power (Phys. § 137) is only perceptible in perfectly transparent crystals. It varies exceedingly in different minerals; thus the precious stones are highly refractive, while other minerals possess this property in a slight degree. What is called *double refraction* is a very peculiar phenomenon. Many minerals not only refract the incident ray, but divide it into two parts, each proceeding in a different direction, so that two images are seen of any object, such as a black line, when viewed in a certain direction through the crystal. Iceland spar is a mineral which exhibits this double refractive power with great distinctness.

15. The *lustre* of minerals is dependent on the nature of their surfaces; and the more these approximate to the reflecting surfaces of mirrors, the more perfect it is. Minute flaws, inequalities, all give rise to certain peculiarities of lustre, which have received various easily intelligible denominations, according to their nature and vividness.

Thus we distinguish: *metallic lustre*, *adamantine lustre*, *vitreous lustre*, *waxy* or *fatty lustre*, *pearly* or *nacreous lustre*, and *silky lustre*. Some minerals are also described as *splendent*, *shining*, *glistening*, *glimmering*, and *dull;* those of earthy fracture, for instance, are distinguished by the latter term.

The *colours* of minerals are expressed by the terms generally adopted for denoting the impressions they produce on the eye. The principal colours are *white*, *grey*, *black*, *blue*, *green*, *yellow*, *red*, and *brown;* besides these there is a variety of mixed colours of all possible shades. A scale of colour, similar to the scale of hardness, has been constructed, the colour of a certain mineral being distinguished by a particular name.

The *streak* of a mineral is also a remarkable characteristic; it is the colour that appears on rubbing or streaking a mineral on a white body, or on scratching it with a harder substance. The streak is, generally speaking, lighter than the colour of the mineral, thus, for example, manganite is nearly black, whilst on paper it produces a brown streak. The colour of a mineral often agrees with that of its streak, yet lively-coloured minerals frequently yield very pale or even colourless powders. The streak of a mineral, that is the colour of the powder produced by scratching it, is a much more trustworthy character than its external colour.

There are many other phenomena of colour, such as *opalescence* and *iridescence*, which, however, occur less frequently. Some minerals under certain conditions, for instance when they are heated, or exposed for some time to the rays of the sun, possess the property of becoming slightly luminous in the dark. This property is called *phosphorescence.*

Relation of Minerals to Electricity and Magnetism.

16. Physics has taught us (§ 151) that all bodies are classed into two groups, one comprising those that become electrical by friction, which are called *electric bodies*, the other formed of those bodies that do not exhibit this property, and which are therefore termed *non-electric.* The electric bodies are non-conductors, and the non-electric bodies are conductors of electricity. It may readily be ascertained to which group a mineral belongs, by rubbing it, and then approaching it to an electrometer. Generally speaking, those minerals that contain heavy metals belong to the class of non-electrics or conductors, while minerals that consist of the non-metallic elements, and the compounds of the lighter metals, become electric by friction, and are non-conductors or imperfect conductors of electricity.

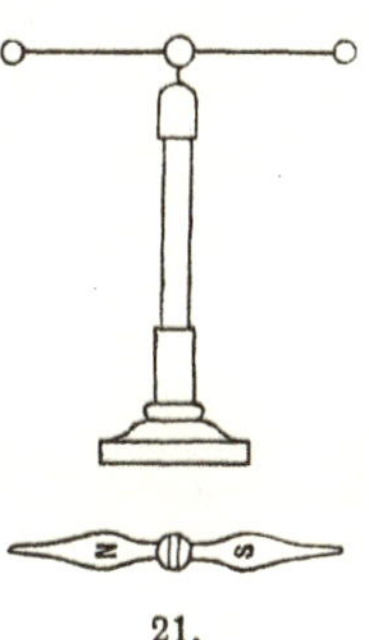
21.

Comparatively few minerals, and only those which contain iron (comp. Phys. § 168) exhibit *magnetic* properties. The magnetic deportment of a mineral may easily be detected by placing it in juxtaposition to a magnetic needle. Fig. 21 represents an electrometer mounted on a point. Below it is a magnetic needle adapted to the same support.

Odour, Taste, and Feel of Minerals.

17. Most minerals have no *odour*. Some have a very characteristic odour, which generally arises from the presence of foreign substances, particularly of mineral naphtha (Chem. § 170); sometimes, however, the odour becomes perceptible only on striking, rubbing, or breathing on the mineral. Several minerals, such as those containing sulphur and arsenic, evolve peculiar odours when heated, these odours being due to chemical changes.

Taste is only possessed by such minerals as are soluble in water, of which there are comparatively few. Its nature is of course dependent on the composition of the mineral; thus the taste of rock-salt is *saline*, that of salts of magnesia *bitter*, that of nitrates *cooling*, &c.

The *touch* or perception experienced on handling many minerals is very characteristic, some being *rough*, as lava; others *fatty* or *unctuous*, as soapstone or talc, whilst others, such as the precious stones, produce a sensation of *cold* when touched. Several minerals possess the property of absorbing water, and some with such power as to adhere firmly to the tongue or a moistened finger, when brought in contact with it; clays exhibit this property in a remarkable degree.

3. CHEMICAL PROPERTIES OF MINERALS.

18. As we have already described minerals as chemical compounds occurring in Nature, they must necessarily be possessed of properties bearing a certain relation to their constituents, and becoming particularly evident when the minerals are decomposed. Chemical formulæ are also employed with great convenience in distinguishing those minerals which have a definite chemical composition. It is therefore of advantage to have become already acquainted with Chemistry, since in Mineralogy we have to refer to that science at every step.

When structure and physical characteristics are insufficient to enable us to recognise a mineral or determine its nature, we must have recourse to chemical action. The mineralogist has then to solve two problems by means of Chemistry; first, the *nature* of the substances contained in the mineral, and then the *quantity* in which each substance is present. To arrive at a knowledge of the latter it is necessary to effect a perfect separation of the mineral into its constituents, and accurately to weigh them. This operation is called *quantitative analysis*, and requires much time and care.

Qualitative analysis merely makes us acquainted with the various substances contained in a body, and, generally speaking, may be conducted with greater expedition, particularly by the mineralogist, who has other auxiliary means at his disposal, for recognising a mineral. He therefore confines himself, as far as possible, to the most simple chemical agents, which he can easily take about with him, and have continually at command. He avails himself especially of the decomposing power of heat, and the solvent property of water and acids. The submission of a mineral to the former is termed its investigation in the *dry way*, and to the latter its examination by the *moist way*.

Action of Heat on Minerals.

19. The Mineralogist applies heat in various degrees of intensity, from gentle

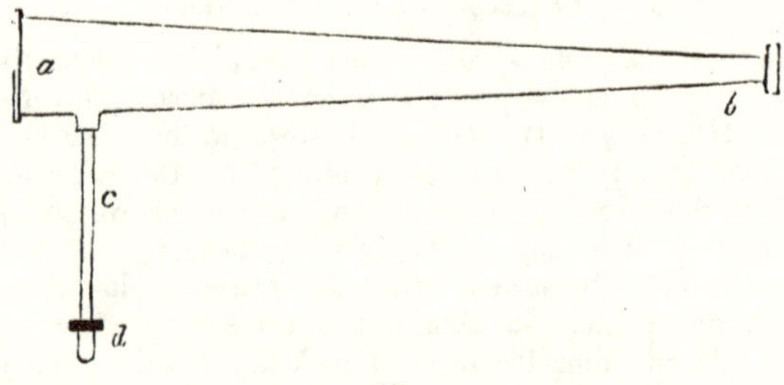

22.

warming to the most powerful ignition. For the latter purpose he makes use of the *blowpipe* (fig. 22), which is a tube of metal, *a b*, terminating in a point with a narrow orifice *d*. The end *b* is the mouth-piece. The blowpipe is made 7 or 8 inches in length, according to the eyesight of the experimenter. On forcing, by means of the blowpipe, a jet of air into the flame of a lamp

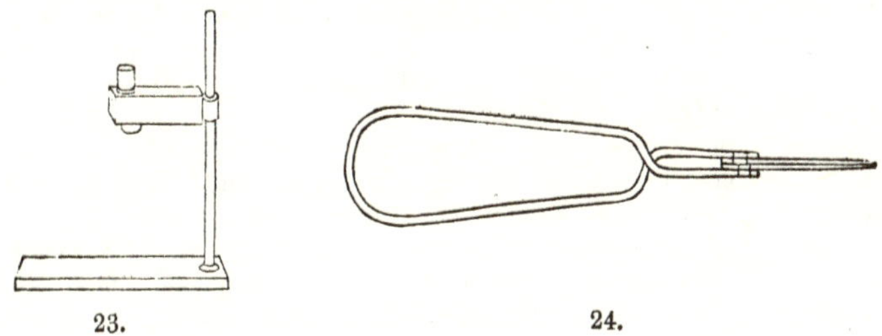

23. 24.

(fig. 23), we obtain on a small scale the same effect which the smith produces by the bellows of the forge, namely, an intense heat in a confined space. The blowpipe imparts to the flame a conical form; into this flame small fragments of the mineral to be examined are introduced, being either held by a small pair of forceps with platinum points (fig. 24), or placed upon a piece of well-burnt charcoal, or a charcoal pastile (fig. 25).[1] When the specimen is to be only gently heated, it is frequently heated in a glass tube by means of a spirit lamp without the aid of the blowpipe.

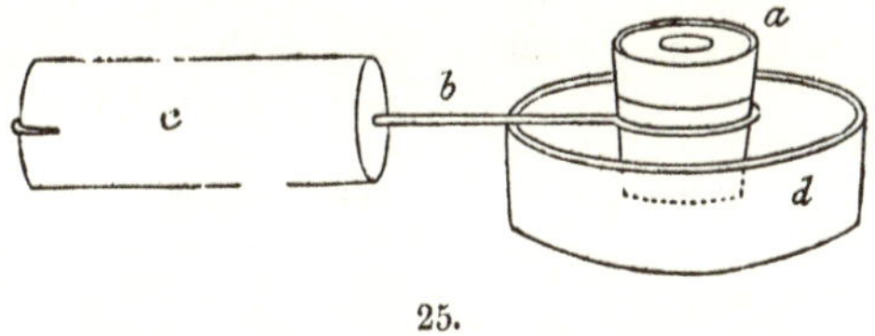

25.

20. In performing these experiments, the particular points to which attention is directed, are the fusibility and volatility of the substance, as also the particular colour which it may impart to the blowpipe flame.

The *fusibility* of minerals varies exceedingly. Some fuse at a gentle heat in the ordinary flame of a lamp, as salts, for instance; others are fused only by the application of the most intense heat; and others again are perfectly infusible. The different degrees of fusibility are expressed by the terms easily fusible, difficultly fusible, infusible, &c.

The fusion of substances is attended by other phenomena worthy of notice; thus some minerals fuse quietly, others swell up or intumesce, or decrepitate, &c. The fused mass presents itself in the form of a glass, a slag, a scoria, an enamel, or a metallic bead, the latter being generally formed when a heavy metal is present.

Volatile substances are very often expelled from minerals by the application of heat. Thus aqueous vapour is almost always evolved, and it is necessary to observe whether this water be merely accidental, or chemically combined as water of hydration or of crystallisation (Chem. § 28). Many minerals disen-

[1] Fig. 25, *a*, represents a charcoal pastile of the full size; *b*, an iron-wire handle; *c*, a cylinder of cork; *d*, a porcelain capsule. The lower part of the pastile is chiefly of clay, and is incombustible; the upper part of charcoal. Another variety of blowpipe pastile contains soda and borax among the charcoal, and is used when metallic ores are to be reduced.

gage various gases, as, for instance, chalk evolves carbonic acid, and binoxide of manganese evolves oxygen. New compounds are often produced during ignition, by the combined action of heat and the oxygen of the air. Thus, lead ores become easily coated with a yellow crust of oxide of lead, ores of antimony with white oxide of antimony, sulphuretted ores yield sulphurous acid, which is easily recognised by its suffocating odour, and arsenical ores evolve the peculiar garlic odour of arsenic.

The *colour of the blowpipe flame* is often an excellent means of distinguishing a mineral. Strontia imparts to the flame a purple tint, lime a bright red, potassa a violet, soda a bright-yellow, boron and copper a green tint, &c.

21. We have hitherto spoken of the test examination by the blowpipe flame alone. The co-operation of chemical substances is frequently employed, by which peculiar phenomena are produced. Such substances are, the oxygen of the air, the carbon of the *interior* of the blowpipe flame, carbonate of soda, and borax or biborate of soda.

We have already seen in § 20 that the oxygen of the air exerts an oxidising influence, and it must here be remarked, that the point of the blowpipe flame is the only portion which allows the oxygen to have access to the substance, it is therefore called the *oxidising flame* of the blowpipe. If the substance under examination is introduced into the wide interior portion of the flame, which is not luminous, and still contains unconsumed carbon, the latter exercises a reducing action, if the substance contains an oxygen compound. This portion of the flame is hence called the *inner* or *reducing flame*. Thus, for instance, a piece of tin may easily be converted into white oxide in the outer flame, and reduced again to a metallic bead in the inner flame.

22. Carbonate of soda and borax, when added to the specimen under examination before the blowpipe flame, are called *fluxes*, since they produce easily fusible compounds. Carbonate of soda when fused with compounds rich in silica, forms an easily fusible soda-glass; it likewise serves for the production of soluble salts of arsenic, sulphur, manganese, &c., which are converted into acids by exposure to a high temperature. In borax (Chem. § 62), the boracic acid that withstands the action of fire, combines with metallic oxides, forming peculiarly coloured glasses, which correspond pretty well in their colours with those of the glass fluxes with which we have previously become acquainted (Chem. § 77). The result obtained in this experiment is dependent on the part of the flame in which the fusion is effected, since the lower oxides frequently yield glasses which differ in colour from those produced by the higher oxides, as is shown by the following examples (see page 281).

The mixture of borax and the substance to be tried is melted on the end of a fine platinum wire, bent into a ring of one-eighth of an inch in diameter.

Similar coloured beads are produced by fusing metallic bodies in a glass formed of microcosmic salt (the double phosphate of soda and ammonia). A crystal of this salt is fused on a charcoal pastile, fig. 26, till it has the form of fig. 27. It is then mixed with the metallic body and again fused, and the heat is continued till the charcoal capsule burns away and leaves the bead exposed, as shown in fig. 28.

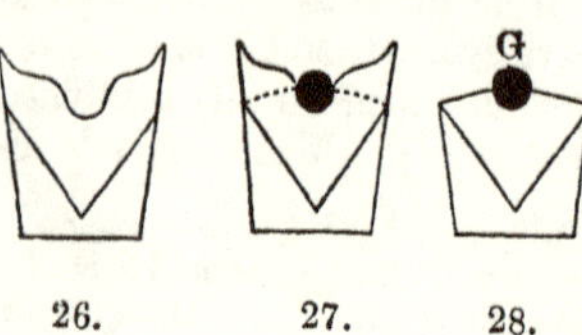

26. 27. 28.

Oxides.	Colour of the Borax Glasses.	
	In the Oxidising Flame.	In the Reducing Flame.
Chromium . .	Emerald-green	Yellowish-brown, colourless on cooling.
Manganese . .	Violet	Colourless.
Antimony . .	Bright-yellow	Turbid and greyish.
Bismuth . . .	Colourless	Grey and turbid.
Zinc	Colourless, a white enamel with much zinc	. . .
Tin	Colourless	Colourless.
Lead	Yellow, colourless on cooling .	Reduced to metallic beads.
Iron	Dark red, becoming lighter and nearly colourless on cooling .	Bottle-green, blue-green.
Cobalt . . .	Blue	Blue.
Nickel . . .	Reddish-yellow, lighter on cooling	Greyish.
Copper . . .	Blue	Colourless, cinnabar-red, and opaque on cooling.
Silver . . .	Milky white on cooling . . .	Greyish.

23. In finally availing ourselves of the co-operation of water and acids as solvents of minerals, we enter at once into the range of those chemical phenomena, which are followed out in all their details in special works on analytical chemistry.

It therefore only remains to be observed, that these solvents are generally applied in a certain order; namely, water first, then hydrochloric acid, then nitric acid, and finally a mixture of the two (Chem. § 36). Hydrochloric acid is most frequently employed, in order to observe whether the minerals effervesce; that is, whether they contain carbonic acid, which escapes as gas when the mineral is put into this acid.

24. We have now made ourselves acquainted with all the preliminary knowledge required to enable us to proceed to describe Minerals. It must, however, be remarked, that of all sciences Mineralogy is the one in which mere description, even of the best kind, is least available. In this science self-inspection is absolutely necessary. The object proposed is not a purely theoretical knowledge of minerals, but a practical acquaintance, only attainable through the medium of our senses; a facility in combining into one conception all the different qualities of such objects, whereby the characteristics are permanently, indissolubly, and unceasingly associated in our minds with the bodies which are so characterised.

It is, therefore, advisable that the student who intends to engage in the pursuit of Mineralogy, should avail himself of those minerals which are furnished by the neighbourhood in which he is living. Even the poorest districts have some minerals, and the examination of these will aid him in forming a conception of others. It is by no means difficult to obtain gradually the most important minerals by exchange or purchase, and thus to form a small collection. Small systematic collections of specimens of minerals can now be purchased for a trifling sum. In all institutions where this branch of natural science is embraced in the course of study, it is necessary above all things to excite an interest for the science by the aid of a collection of the most important minerals. In the study of natural history, the best description may be considered merely as a crutch, which is cast aside directly the student has an opportunity of inspecting the objects personally.

Classification of Minerals.

25. A mineral which may be distinguished from all others by its peculiar chemical composition and properties, is acknowledged as a distinct *Species.* The number of minerals thus established amounts to four or five hundred, and is rapidly increasing.

Minerals may be arranged according to various systems. Either their form is principally considered and they are then arranged according to the systems of crystals, or their arrangement is based upon their density and hardness. Since, however, it has been more clearly shown that all these properties are dependent on the chemical constitution of the minerals, the latter has become the directing clue to their arrangement. In this classification particular regard is paid to that constituent which either predominates in quantity, or in its particular character, and which therefore furnishes the name for the formation of the group. The order of succession of the minerals in this arrangement, is nearly the same as that of the elements and their compounds in chemistry, although gaps are found to exist here and there in the system.

The acquirement of a knowledge of chemistry is of course presupposed, whereby a number of difficulties will vanish, which would render the study of mineralogy, according to outward characteristics alone, exceedingly laborious.

26. The nomenclature of minerals has been formed gradually, without any scientific basis, and is consequently imperfect. The names of genera and species are derived from many sources; as, for example, from popular or vulgar names, from the locality where they were first noticed, and from the names of celebrated naturalists; and but few names have been derived from their properties and chemical constituents. An alteration in the nomenclature cannot however be effected, as it would give rise to the greatest confusion. The old names are retained as a matter of convenience, just as, in chemistry, the names water and potash are still employed, instead of the more systematic names of oxide of hydrogen and oxide of potassium.

4. DESCRIPTION OF MINERALS.

27. A considerable space would be required for the description of all the minerals that are now known. We must therefore content ourselves with describing the most important, and even these only briefly. A sufficiently detailed account has been given of several; for instance, of the different kinds of coal, in the chemical section of this work: of these, therefore, the mere enumeration will suffice.

Most of the simple minerals occur in comparatively small quantities, though some which are aggregated in large masses form a considerable portion of the earth's surface. These will be referred to in the chapter on Rocks.

In the following descriptions, H. signifies the hardness, and Sp. Gr. the specific gravity, of the minerals:—

SYNOPTICAL TABLE OF MINERALS.

1st CLASS. NON-METALLIC BODIES.	2nd CLASS. METALS.		3rd CLASS. ORGANIC COMPOUNDS.
	1st Order. Light Metals.	2nd Order. Heavy Metals.	
Group.	Group.	Group.	Group.
1. Sulphur.	5. Potassium.	13. Iron.	29. Salts.
2. Boron.	6. Sodium.	14. Manganese.	30. Earthy resins.
3. Carbon.	7. Ammonium.	15. Cobalt.	
4. Silicium.	8. Calcium.	16. Nickel.	
	9. Barium.	17. Copper.	
	10. Strontium.	18. Bismuth.	
	11. Magnesium.	19. Lead.	
	12. Aluminum.	20. Tin.	
		21. Zinc.	
		22. Chromium.	
		23. Antimony.	
		24. Arsenic.	
		25. Mercury.	
		26. Silver.	
		27. Gold.	
		28. Platinum.	

FIRST CLASS.—MINERALS OF THE NON-METALLIC ELEMENTS.

1ST GROUP—SULPHUR.

28. The primary form of crystallised Sulphur is the rhombic octohedron occurring with various truncations of the edges and angles (figs. 29, 30). Sulphur occurs more frequently in the semi-crystalline or granular, and the earthy state, and less frequently in the fibrous condition. Its cleavage is imperfect; its fracture conchoidal and uneven; H = 1·5 to 2·5; it is brittle and fragile; Sp. Gr. = 1·9 to 2·1. The chemical properties of sulphur and its application have been described in the section Chemistry (§ 40).

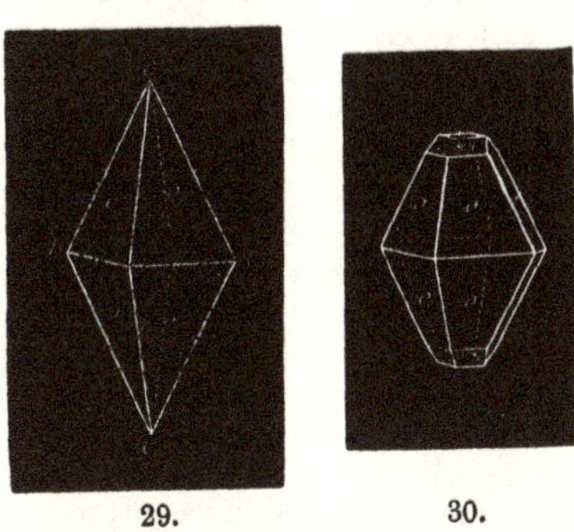

29. 30.

The most important locality of sulphur is Sicily, where it is found in the tertiary formations, associated with calcareous spar and celestine; at Girgenti, Fiume, &c. Moreover, the strata of earthy sulphur in Poland are likewise very considerable. In addition to these localities, there are many places in Germany and the rest of Europe, as also in other parts of the globe, where sulphur is found. But the sulphur derived from these sources is far less pure than the sulphur of Sicily.

2ND GROUP—BORON.

29. This element occurs only in combination with oxygen, as Boracic acid (BO_3+HO), which crystallises in scales, and is found as a crust upon the surface of the earth in the neighbourhood of volcanic springs. It is friable, of Sp. Gr. =1·48, translucent, white, bitter and acid to the taste; it is easily fusible, and imparts a green colour to flame; it is soluble in water and alcohol. Boracic acid is deposited at the margin and at the bottoms of volcanic springs or lakes, particularly in those of Sasso (hence the name Sassolin), Castelnuovo, and others, in Tuscany.

3RD GROUP—CARBON.

30. (1.) *Diamond.*—This mineral occurs crystallised in regular octohedrons, or in some of the geometrically allied forms (figs. 31, 32, 33). It is possessed

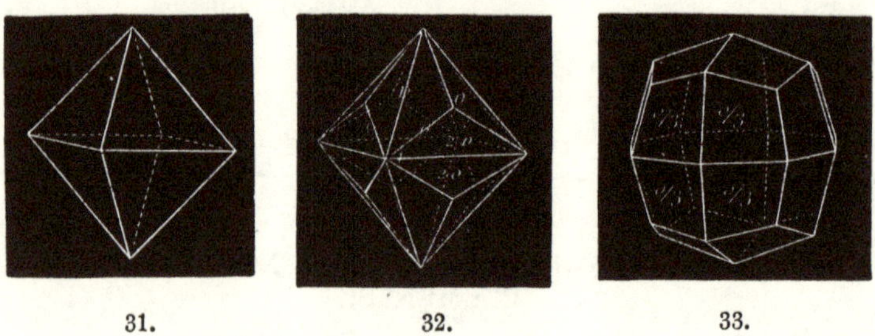

31. 32. 33.

of the greatest hardness, which is = 10; Sp. Gr. = 3·5 to 3·6; it is mostly cleavable, transparent, generally colourless, highly lustrous, refracts light very powerfully, and is the most valuable of all the precious stones. It occurs principally in alluvial soil and in rocks of secondary formation, in the East Indies (Golconda), and in Brazil. It has also lately been discovered in the sands of the rivers which have their sources in the Uralian mountains: 1 carat (= 4 grains) of small diamonds, employed for polishing the larger ones, for cutting glass, &c., costs from 20*s.* to 25*s.* A polished diamond (brilliant) weighing 1 carat, is valued at from 8*l.* to 10*l.*; the prices of diamonds increase to such an extent with their size, that a brilliant weighing 5 carats may cost as much as from 150*l.* to 250*l.* The largest diamond at present in Europe is one in the possession of the Queen of Portugal. It weighs 215 carats, and is valued at upwards of 150,000*l.*

(2.) *Graphite* (Plumbago) is found in tabular crystals, belonging to the hexagonal system, but generally it occurs in scales and small laminæ. H = 1 to 2, Sp. Gr. = 1·8 to 2·4; it is cleavable, steel-grey to black, unctuous to the touch, and produces a black streak. It is found embedded in various rocks at Passau, in Bavaria; the finest quality, however, is met with at Borrowdale, in Cumberland. The graphite from the former place is generally employed for crucibles and for blacking stoves, and that from the latter locality for the best black-lead pencils.

(3.) *Anthracite* occurs in large masses, having a conchoidal fracture. H = 2 to 2·5, Sp. Gr. = 1·4 to 1·7; it is greyish-black, and leaves but little ash when burned. It is found in strata, occasionally of very considerable thickness, in the primitive rocks; as, for instance, in Wales and in the Hartz mountains. It is employed as a fuel for strong blast or wind furnaces, &c.

Coal, *brown-coal*, and *turf* must be mentioned as varieties of carbon, since this element forms their chief constituent. Their most important characters and properties have already been detailed in § 164 of Chemistry. We shall, however, more minutely describe the nature of their stratification in the chapter on Geology.

4th Group—SILICIUM.

31. By the term silica the Mineralogist recognises the compound, which chemists call silicic acid (SiO_3, Chem. § 61). The number of siliceous minerals is exceedingly large; silica, however, generally occurs in combination with alumina, hence the greater number of silica compounds is mentioned in the alumina group. It may in general be remarked, that the hardness of the purer kinds of silica is very considerable, sometimes amounting to 8·9; hence it produces sparks when struck with steel, whilst its specific gravity rarely exceeds 4·5. It possesses mostly a vitreous lustre, its prevailing colour being white. Silica when chemically pure, or merely coloured by small quantities of different oxides, is termed *quartz*.

Quartz Family.

Its crystals belong to the hexagonal system, and occur most frequently as double six-sided pyramids (fig. 18), but which are generally subjected to the modifications and irregularities spoken of in the article on Crystallography, and in part represented in the figures 34 to 37. Quartz also occurs frequently in crys-

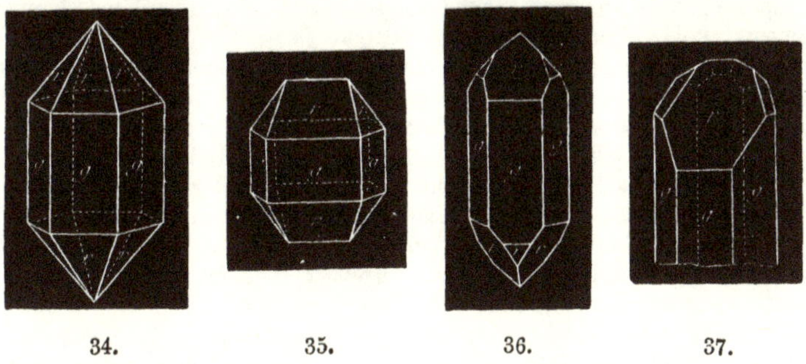

34. 35. 36. 37.

talline, compact, or granular masses. Its fracture is conchoidal; H = 7; Sp. Gr. = 2·5 to 2·8. It is either perfectly transparent, or white; it occurs likewise of all colours, and of every possible shade. It is insoluble in all acids, with the exception of hydrofluoric acid (Chem. § 39); it yields with carbonate of soda, before the blowpipe, a transparent *glass;* and when struck with steel it emits brilliant sparks. The following are the principal varieties of quartz:—

(1.) *Rock crystal*, which is found in beautiful, transparent, six-sided prisms of considerable size, in the various primitive mountains. The crystals obtained from the caverns of St. Gotthardt are remarkably fine; they have also been found of extraordinary size and purity in Madagascar, where blocks of from 15 to 20 feet in circumference occur. Rock crystal is employed in jewellery, and as a constituent of pure glass fluxes. It is often slightly coloured, and frequently contains various foreign minerals, either in small scales or in other forms.

(2.) *Amethyst* is quartz that is more or less intensely coloured by protoxide of manganese. It occurs rarely in perfectly formed crystals, but more frequently as crystals lining drusic cavities. It is often found in the cavities of porphyry and amygdaloid, and, as it is by no means a rare mineral, is extensively employed as a jewel of small value. The amethyst was worn by the ancients as a charm against drunkenness.

(3.) Silica is called *common quartz*, when it does not occur in regular crystals, but only crystalline, granular, and compact. In this state it is often found in considerable masses, which are called quartz rock; it also forms compound rocks with other minerals, of which granite offers a familiar example. It is very extensively dispersed over the surface of the earth. Its purer kinds are employed in the manufacture of glass, porcelain, &c. It is generally white and translucent; some varieties of it, altered in colour or otherwise, have received different names, as *rose quartz;* the blue variety, *siderite; schiller-spar; cats-eye*, so called from its peculiar iridescence; *avanturine*, containing yellow and reddish laminæ of mica, which render it a very beautiful and ornamental stone.

(4.) *Calcedony* is an opaque kind of quartz, occurring in spherical, botryoidal, and nodular masses, possessing the most varied colours and curious markings. It is employed extensively for making snuff-boxes, buttons, marbles, &c. Red or yellow-coloured calcedony is called *carnelian*, and the green-coloured, chrysoprase; both are much prized for seals, and other raised and engraved ornamental works of art.

(5.) *Flint*, the properties of which are well known, is found in large irregular masses, in many parts of England and about Paris. Its application as a promethean apparatus has diminished considerably since the invention of lucifer-matches and percussion-caps. It is extensively used in potteries.

(6.) *Hornstone* is a variety of quartz somewhat similar to flint, but has a more splintery fracture, and bears a remarkable resemblance to horn.

(7.) *Jasper* is opaque, dull, and only slightly lustrous, on account of the larger amount of alumina and oxide of iron which it contains. It occurs of all colours; but red, yellow, and brown are most frequent.

(8.) *Flinty slate* is a mineral consisting of quartz, alumina, lime, and sesquioxide of iron, coloured black by carbon; it is employed as a whet-stone and touch-stone.

(9.) *Agate* is, generally speaking, a beautifully-marked mineral, consisting of a mixture of various kinds of quartz, particularly of amethyst, calcedony, and jasper. It is extensively employed by lapidaries for making a variety of ornamental objects, and also for mortars, which are employed for pulverising very hard substances.

OPAL.

32. This mineral is a particular variety of quartz, containing water in chemical combination. It does not occur crystallised, but in compact vitreous masses, and is distinguished by the brilliant and changeable reflections of light exhibited by some of its varieties; whence the term *opalescent* is derived. *Noble opal* possesses this property in a very high degree, and is therefore much prized as a jewel. *Semi-opal*, or *common opal*, is less remarkable for its change of colours. *Hydrophane* is a mineral, possessing the peculiar property of becoming transparent and iridescent only when moistened with water.

Siliceous sinter and *mountain meal* are likewise varieties of quartz, containing water; the former is deposited in a variety of forms by hot springs, particularly by the Geyser of Iceland. The latter is an earthy deposit from siliceous waters, and, when examined under the microscope, is found to consist almost entirely of the shells of infusoriæ. Another kind is called *polishing slate*, and is employed by lapidaries for polishing stones.

SECOND CLASS.—MINERALS CONTAINING METALS.

FIRST ORDER—LIGHT METALS.

5TH GROUP—POTASSIUM.

33. The most important and indeed the greater number of minerals containing potassium, likewise contains alumina, as an essential constituent: we shall therefore describe them in the aluminous group. Of natural potassa salts, we have only to mention the nitrate and the sulphate of potassa.

Nitrate of potassa, or *nitre* (KO,NO_5), crystallises in regular rhombic prisms, but is found, in many localities, in the form of crusts of acicular and capillary crystals (comp. Chem. § 69).

Sulphate of potassa (KO,SO_3) belongs to the same crystalline system, and is found occasionally in volcanic lava.

6TH GROUP—SODIUM.

34. (1.) *Nitrate of soda* (NaO,NO_5) crystallises in the hexagonal system as obtuse rhombohedrons, and occurs in crystalline masses of considerable magnitude; which, in the districts of Atakama and Tarapaca, in Peru, extend over a space of nearly 200 miles.

(2.) *Rock-salt*, chloride of sodium ($NaCl$), crystallises in the cubical system; it generally occurs, however, in tabular crystalline masses, and is easily cleavable in a direction parallel to the planes of the primary form; its fracture is conchoidal, $H = 2$; Sp. Gr. $= 2{\cdot}2$ to $2{\cdot}3$; its colour is generally white, but it is also found of a red, green, yellow, and blue colour; its chemical properties and applications are detailed in Chemistry § 72. Rock-salt occurs in secondary Rocks, in masses of considerable magnitude, often in company with gypsum, alumina, and saline clay. The salt-works of Cheshire, of Hallein in Saltzburg, and of Wielizka, in Gallicia, are particularly celebrated: in the latter is found the *decrepitating salt*, which dissolves in water with a decrepitating noise and disengagement of numerous bubbles of hydrogen gas. The gas is enclosed between the crystalline planes of the salt. A number of other minerals containing soda are found, but are of less importance than the foregoing. Of these may be mentioned anhydrous and hydrated sulphate of soda (Thenardite $=NaO,SO_3$, and Glauberite $=NaO,SO_3+10HO$); carbonate of soda, containing a large quantity of water (Soda $=NaO,CO_2+10HO$), and another kind containing little water, called *trona* ($2NaO,3CO_2+4HO$), which occurs in the interior of Barbary in considerable quantities, as a crust on the earth, and is applicable to the same purposes as soda.

Biborate of soda ($NaO,2BO_3+10HO$), as a mineral, is called borax or tincal, and is found at the bottom and on the borders of a lake in Thibet.

7TH GROUP—AMMONIA.

35. Since the combinations of ammonia, as we have seen in § 78 of Chemistry, are of a volatile nature, they occur only in inconsiderable masses, though not unfrequently, in the mineral kingdom. They are met with principally as crystalline coatings or crusts; for instance, in cavities and the fissures of lava of active volcanos, in brown-coal works, particularly in the neighbourhood of burning and spent heaps of coals.

8TH GROUP—CALCIUM.

36. This metal forms an extensive group of minerals, which possess a low degree of hardness and density, and are generally of a pure white colour. The most remarkable are—

(1.) *Fluorspar* (CaFl), which crystallises in various forms of the regular system, but most frequently as cubes. It is perfectly cleavable; its fracture is conchoidal; H = 4; Sp. Gr. = 3·1 to 3·17; it is transparent and gradually passes into translucency, and seldom occurs white, being generally tinted faintly violet, green, yellow, &c.: regarding its chemical properties, see Chemistry, § 39. Fluorspar is a mineral of frequent occurrence, though never in considerable masses. The same mineral occurs amorphous, as *compact fluor* and *earthy fluor*.

(2.) *Anhydrite* (CaO,SO_3), or anhydrous sulphate of lime, is found generally associated with gypsum and rock-salt. It occurs crystallised, and also in radiated, granular, and compact masses.

(3.) *Gypsum* (CaO,SO_3+2HO), or hydrated sulphate of lime, occurs most frequently in tabular crystals, which may be cleaved into very thin laminæ; they belong to the system of the fourth primary form (fig. 16); H = 2; Sp. Gr. = 2 to 2·4. It frequently occurs in double or twin crystals, of the form represented in fig. 38. It is possessed of double refractive power, vitreous lustre, and generally has a white colour. This kind of gypsum is called *selenite;* there are, besides, other varieties, viz., *fibrous* gypsum; *compact* or *granular* gypsum, which is called *alabaster*, and *earthy* gypsum. Regarding its application, see Chemistry, § 81.

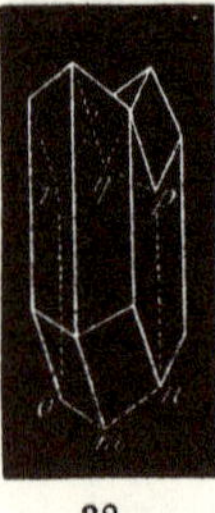

38.

(4.) *Apatite*, sometimes called asparagus-stone, on account of its beautiful pale-green colour, consists of phosphate of lime, fluoride and chloride of calcium. It crystallises in the hexagonal system, often like fig. 41, and is frequently embedded in various kinds of rocks.

(5.) *Pharmacolite* is arseniate of lime ($=CaO,AsO_5$).

(6.) CARBONATE OF LIME (CaO,CO_2).

37. This mineral exhibits the remarkable peculiarity of crystallising in forms belonging to two different systems; hence its varieties form two families, namely, those of Calcareous spar and Arragonite.

(1.) *Calcareous spar* crystallises in the hexagonal system, and more particularly in modifications of the rhombohedron (fig. 20). These are so exceedingly numerous, that 1,700 different forms have already been distinguished. Figures 39 to 44 represent some of the chief forms of carbonate of

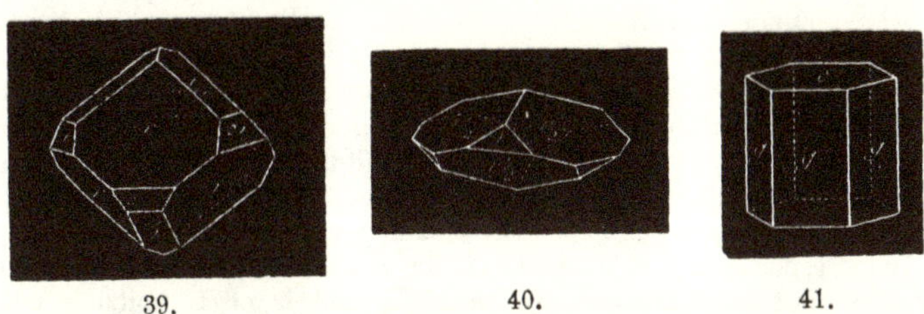

39. 40. 41.

lime, which we shall briefly describe, as they give much insight into the mysteries of crystallography. Fig. 39, *r* shows the planes of the primitive rhombohedron (obtainable from *all* the crystals of calcareous spar by cleavage), combined with $r\frac{1}{2}$, a more obtuse rhombohedron, and $r\,2$, a more acute rhombohedron. In fig. 40, the primary rhombohedron, *r*, is almost entirely obliterated by the large planes of the obtuse rhombohedron, $r\frac{1}{2}$. In fig. 41 we have a regular six-sided prism, such as forms the middle portion of

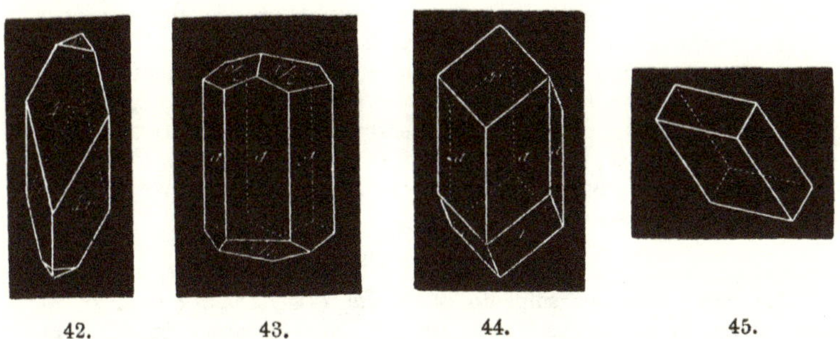

42. 43. 44. 45.

the crystal represented by fig. 19. Fig. 42 represents small faces of the primitive rhombohedron, *r*, combined with a predominant and very acute rhombohedron, $r\,4$. In fig. 43 we have the obtuse rhombohedron, $r\frac{1}{2}$, terminating a regular six-sided prism, *g* (fig. 41), and in fig. 44 the inverse, but equal six-sided prism, *a*, terminated by the regular rhombohedron, *r*. Crystals of all these forms are readily procured among the minerals of Cumberland and Derbyshire.

Fortunately the remaining properties of calcareous spar are such as to admit of its easy recognition. It is very easily cleavable into the primitive rhombohedron (fig. 20), and into tables such as shown by fig. 45. It has a conchoidal, splintery, and uneven fracture; H = 3; Sp. Gr. = 2·6 to 2·17;

it becomes electric by friction, is soluble in the mineral acids with evolution of carbonic acid, and is converted by exposure to a red-heat, into caustic lime (Chemistry, § 79). Its varieties are:—

(*a.*) *Crystallised calcareous spar*, called also Iceland spar, or *double refracting spar*, because it it possessed of double refractive power in a high degree (§ 14). It most generally forms tabular, transparent, colourless crystals of vitreous lustre, often occurring in all varieties of formations, particularly in drusic cavities and in metallic veins. The double refracting spar found in Iceland is celebrated for its beauty. (*b.*) *Fibrous limestone*, which occurs principally in stalactitic formations, in the cavities of chalk-hills. (*c.*) *Marble*, or granular carbonate of lime, which is most esteemed when it is perfectly white, fine-grained, compact, and free from coloured veins. It is employed in that state by sculptors, for their most beautiful productions. The most celebrated marble quarries are those of Carrara, in Italy, and of Paros, in Greece. Coloured marble, frequently containing variegated spots and veins, is much more common; it is employed for pedestals, columns, &c.; it is one of the most beautiful building materials, and is often imitated with coloured and polished gypsum (stucco). (*d.*) *Schiefer-spar.* (*e.*) *Aphrite*, or *earth-foam.* (*f.*) *Compact limestone*, in which no crystalline structure is perceptible, and which generally occurs in large masses, forming entire limestone-hills. It is found in all later mountain formations, in the most varied forms and colours, as *stinkstone*, *marl*, *oolite* (*Roestone*), *calcareous tufa*, &c. (*g.*) *Chalk*, well known as the fine earthy white writing material, occurring in masses in the tertiary formations, particularly in England and in France (Champaigne).

(2.) *Arragonite*, the crystals of which belong to the rhombic system (fig. 13), and generally occur as rhombic prisms. The crystals are sometimes isolated, and often grown together, groups being thus frequently produced, which have the appearance of hexagonal prisms. Arragonite is cleavable; its fracture is conchoidal and uneven; H = 3 to 4; Sp. Gr. = 2·9 to 3; it is transparent, vitreous, and colourless. It is not unfrequently found in the vesicular cavities of basalt and other rocks. It occurs in groups of hexagonal prisms in Arragonia, whence its name is derived. Besides crystallised arragonite, we also find radiated and fibrous arragonite.

9TH GROUP—BARIUM.

38. (1.) *Heavy spar*, or sulphate of baryta (BaO,SO_3), crystallises in rhombic prisms, of which there exist about 73 modifications: fig. 46 represents one of the tabular forms in which this mineral crystallises. It is perfectly cleavable, and exhibits an imperfect conchoidal fracture; H = 3 to 3·5; Sp. Gr. = 4·3 to 4·58, whereby it is easily distinguishable from spathic minerals of nearly similar forms. It is transparent, and possesses double refractive power and vitreous lustre; it imparts a green colour to the blowpipe flame. A piece of heavy spar, when warmed, or heated to redness, will remain luminous in the dark for some time afterwards.

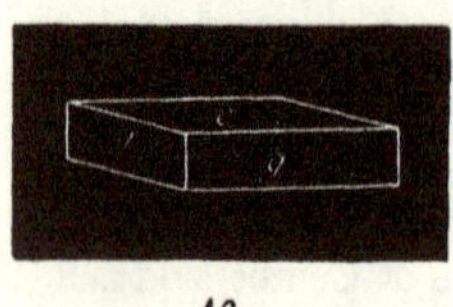

46.

Crystallised *heavy spar* is a mineral of frequent occurrence in mineral veins. It is employed as a white paint, and used to adulterate white-lead (Chemistry,

§ 83). Heavy spar also occurs radiated, fibrous, granular, compact, and earthy.

(2.) *Witherite*, or carbonate of baryta (BaO,CO_2), crystallises in regular rhombic prisms, and is principally found in this country. It is employed as the source of many of the other compounds of baryta; as, for instance, of chloride of barium, nitrate of baryta, &c.

10TH GROUP—STRONTIUM.

39. (1.) *Celestine*, or sulphate of strontia (SrO,SO_3), crystallises in the rhombic system (fig. 13), the rhombic prism being the prevailing form. Its cleavage is perfect; its fracture conchoidal or uneven; H = 3 to 3·5; Sp. Gr. = 3·8 to 3·96; it is transparent, double refractive, colourless or white, of vitreous lustre, and imparts a *crimson colour* to the flame of the blowpipe. It does not occur very frequently. The varieties of this mineral are:—*celestine-spar*, *radiated celestine*, and *fibrous celestine*, which has a blue tint, and is found at Jena; and compact celestine, which contains from 8 to 9 per cent. of carbonate of lime. These minerals are employed for the preparation of strontia-salts (Chemistry, § 84).

(2.) *Strontianite*, or carbonate of strontia (SrO,CO_2), is of less frequent occurrence than the preceding mineral, and crystallises in the same system.

11TH GROUP—MAGNESIUM.

40. This metal forms a rather larger group of minerals than the preceding metals. Among these may be mentioned *periclase*, which is nearly pure magnesia (MgO); hydrate of magnesia (MgO,HO); *boracite*, or phosphate of magnesia; and *hydroboracite*, containing, besides the latter substance, phosphate of lime and water. All these minerals occur but rarely, and in inconsiderable masses. *Sulphate of magnesia* (MgO,SO_3), is of more frequent occurrence; but, on account of its solubility, is only found as thin crusts or films of crystalline fibres in the clefts of rocks. In the Siberian steppes, however, whole districts are found covered with this and other magnesian salts. It is contained in large quantities in magnesian mineral waters, particularly in those of Epsom, Seidlitz, Eger, and Seidschütz.

Magnesite, carbonate of magnesia (MgO,CO_2), occurs either crystallised, as *magnesite spar* (talc-spar), or as compact magnesite. The former crystallises in the hexagonal system, and is found in the form of obtuse rhombohedrons; H = 4; Sp. Gr. = 3. The magnesian limestone, consisting of lime, magnesia, and carbonic acid (CaO,CO_2+MgO,CO_2), is a mineral occurring in larger masses. Its crystalline variety is called *bitter spar*, and sometimes *brown spar*. It occurs in obtuse rhombohedrons, nearly resembling fig. 20, which are easily cleavable, and of conchoidal fracture; H = 3·5 to 4; Sp. Gr. = 2·8 to 3. It is semi-transparent, has a vitreous lustre, and is white, or frequently coloured yellow or brown by the presence of iron or manganese. It is principally found in clefts and cavities of the granular magnesio-calcite, called *Dolomite*. The white crystalline variety of Dolomite resembles marble, the coloured kinds are like common varieties of limestone; and as it occurs in large masses, it is employed for similar purposes.

The combinations of magnesia with silicic acid, form a particular class of minerals, of which *Talc* is a prominent member. This mineral contains 62 per cent of silicic acid, 30 per cent of magnesia, and occurs principally in aggregates of imperfect crystals. It is smooth and unctuous to the touch, is very soft, and is white or faintly coloured. It is found in quantities as chlorite slate. A variety of this mineral, called pot-stone, which may be cut and turned, is employed for the manufacture of various vessels. Besides the above, we shall mention, in connection with magnesia, the serpentine and augite minerals, which may be grouped into families.

1st Family—Serpentine.

41. This class comprises soft minerals, which may be cut with a knife, their hardness rarely exceeding 2·3. They do not occur in crystals, but are mostly opaque, difficultly fusible, and but slightly lustrous. They principally consist of magnesia and silicic acid, generally coloured by oxide of iron. To this family belongs the unctuous Steatite, or soap-stone, which is employed for removing grease-spots, or as a soft polishing powder, as also for the manufacture of a great variety of objects of art. The most common varieties are French chalk, and the well-known meerschaum, which is used for the manufacture of tobacco-pipes. *Serpentine*, which is called ophite, or snake-stone, on account of its green-spotted appearance, resembling the skin of a snake, forms compact masses of granular fracture, occurring as rocks. Its hardness is 3; it is employed for the manufacture of a number of objects, particularly columns, boxes, mixing-mortars for chemists and druggists, &c. There is moreover a large number of minerals resembling serpentine, which may be classed in this family.

2nd Family—Augites.

42. These minerals possess a hardness between 4·5 and 7, and a specific gravity of 2·8 to 3·5. Their prevailing colours are dark-green and black; they are fusible before the blowpipe. Their principal constituents are silica and magnesia, but some of them also contain considerable quantities of other oxides, such as sesquioxide of iron and alumina, which render it difficult to classify these minerals according to their chemical constituents. The augites occur in peculiar crystalline forms, and not unfrequently in considerable masses. They are also contained in many varieties of rocks, such as lava, basalt, &c. The most important members of this family are augite and hornblende, of which the various kinds are again distinguished by different names.

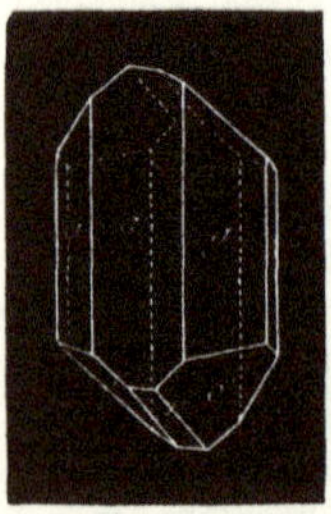

47.

Augite crystallises in prisms belonging to the fourth system; fig. 47 represents one of its usual forms: its different varieties occur principally in volcanic formations and their vicinities. The most notable are—*diopside*, *diallage*, *bronzite*, *hypersthene*, and *kokolite*.

Hornblende belongs to the same crystalline systems as the preceding mineral, with which it likewise exhibits similarity in its chemical composition and colour. *Asbestos*, *amianthus*, and *mountain cork*, must be viewed as varieties of hornblende, crystallised in exceedingly fine needles. The most pliable kinds of asbestos are mixed with flax, and woven into stuffs and cloths, from which the flax may be removed by simple ignition; and thus incombustible cloths are prepared, which may be worn in cases of danger from fire. The dead bodies of the rich were, in ancient times, enveloped in such garments, and then burned; by which means their ashes were preserved.

12TH GROUP—ALUMINUM.

43. This group is exceedingly large and numerous, and must therefore be subdivided into families. Of these minerals there are only a few which contain the sesquioxide of aluminum, or alumina, as the chief constituent. It forms, however, the principal compound when in combination with silicic acid; and the large amount of the latter substance in a mineral, frequently renders it doubtful whether it should not rather be classed among the silicious than the aluminous group. This class contains a large number of minerals, which are important to the arts and to agriculture; it likewise includes the most precious jewels, next in value to the diamond itself.

1ST FAMILY—CORUNDUMS.

44. These minerals, consisting of pure alumina (Al_2O_3), occur in various forms. (1.) The crystallised variety is *sapphire*, which is found in various modifications of the hexagonal system. It is cleavable, and of conchoidal fracture; Sp. Gr. = 4; H. = 9; it is perfectly transparent, possesses a highly vitreous lustre, and beautiful blue colour; it is, however, also found of a red, green, yellow, and white colour, the red variety, which is called *ruby*, being very highly prized. The above properties render sapphire a very valuable gem: it occurs in small crystals in Germany, but the finest specimens are found in the East Indies, in diluvial soils and in the sand of rivers, which have their source in such formations.

(2.) *Common corundum* is found in rough, scarcely translucent, dull, or dirty-coloured crystals, embedded in granitic rocks; being possessed of great hardness, it is reduced to powder, and employed for cutting and polishing other precious stones. (3.) *Emery* occurs in compact or granular masses, which are found in Saxony, in Greece, and in other localities, embedded in mica-slate. It is but slightly lustrous, and has a bluish-grey colour. Its powder is frequently employed for cutting and polishing.

2ND FAMILY—ALUMS.

45. (1.) *Aluminite* (Al_2O_3,SO_3+9HO) is basic sulphate of alumina, and is found in small quantities as a white earthy mass. (2.) *Sulphate of alumina* ($Al_2O_3,3SO_3+18HO$), termed also feather-alum, occurs in fibrous crystalline crusts, or in porous and compact masses. (3.) *Alum-stone*, consisting of alumina, potassa, and sulphuric acid, crystallises in the hexagonal system as rhombohedrons, and is found particularly in the vicinity of Rome, where it is employed for the preparation of Roman alum; which, as it contains no iron in chemical combination, was for a long time particularly prized, until the progress of Chemistry made us acquainted with other methods of preparing alum free from iron. (4.) *Alum* ($KO,SO_3+Al_2O_3,3SO_3+24HO$), with which we have become acquainted in the section Chemistry, § 87, occurs likewise in Nature, crystallised in regular octohedrons. It is an interesting fact, that

various minerals exist which have a composition corresponding to that of alum, in which the potassa is replaced by other bases, without the form of the crystal being in the least altered. Thus we are acquainted with:—

Potassa-alum	=	$KO,SO_3 + Al_2O_3,3SO_3 + 24HO.$
Soda-alum	=	$NaO,SO_3 + Al_2O_3,3SO_3 + 24HO.$
Ammonia-alum	=	$NH_4O,SO_3 + Al_2O_3,3SO_3 + 24HO.$
Manganese-alum	=	$MnO,SO_3 + Al_2O_3,3SO_3 + 24HO.$

a series of compounds, the formulæ of which present the greatest similarity. Such compounds as the above, containing different constituents, but crystallising in the same form, are termed *isomorphous*, that is, of similar form: we shall meet with several other examples of isomorphism as we proceed.

Phosphate of alumina is likewise found in the crystalline form, and is called *Wavellite.*

3RD FAMILY—SPINELS.

46. These minerals are combinations of alumina and magnesia, and are represented by the formula MgO,Al_2O_3, in which the alumina occupies the place of an acid. They crystallise in regular octohedrons and in modifications of this form; they are distinguished by their hardness (H. = 8; Sp. Gr. = 3·8), lustre, and transparency, and are prized as valuable gems. Various kinds of spinel are distinguished by the colour: the scarlet variety, which is called *spinel ruby*, is the most highly prized; it occurs in the East Indies. Besides this variety, we are acquainted with blue, green, and black spinels.

4TH FAMILY—ZEOLITES.

47. The *Zeolites*, or *boiling stones*, so called on account of their containing water, with which, when heated before the blowpipe, they part under *intumescence*, are mostly white, vitreous, and transparent; they possess a hardness of 3·5 to 3·6, and a specific gravity of from 2 to 3. Their principal constituent is silicate of alumina, which, in the different varieties, is combined with variable quantities of silicate of potassa, soda, and lime, and with water of crystallisation often in considerable quantities. Although these minerals are interesting on account of their chemical composition, and particularly the variety and peculiarity of their crystalline forms, there is no member of the family that is of any importance with regard to frequency of occurrence or technical application. We must confine ourselves to mentioning a few of the best known zeolites, such as *analcime*, *harmotome*, or cross-stone, so called from the crystals often crossing each other at right angles, *stilbite*, *chabasite*, *mesotype*, *natrolite*, *prehnite*, *Thomsonite*, &c. Very beautiful specimens of all these minerals are found in cavities in the basaltic rocks near Kilpatrick and Kilmacolm, on the Clyde.

5TH FAMILY—CLAYS.

48. By the term *clay* is understood a chemical combination of silica with alumina (Al_2O_3,SiO_3), as has already been mentioned in Chemistry, § 87. The minerals of which clay is the principal constituent are either crystallised, possessing a hardness of about 7·5, transparent, and of vitreous lustre, or they

are compact or earthy. All varieties of clay are difficultly fusible or perfectly infusible before the blowpipe. The more remarkable are :—

(1.) *Andalusite*, which occurs in regular rhombic prisms: H. = 7·5; Sp. Gr. = 3·1 to 3·2: it is infusible and generally flesh-coloured. (2.) *Chiastolite*, so called (from the Greek) in allusion to its being marked with the Greek letter chi (X), visible on the cross sections of the crystals. (3.) *Disthène*, which crystallises in columns, belongs to the 4th system, and acquires a bluish luminosity when gently heated: H. = 5 to 7; Sp. Gr. = 3·5 to 3·6.

The following are earthy clays, coloured red, yellow, or brown by sesquioxide of iron or its hydrate: *Yellow ochre*, which is used as a colour. *Tripoli*, employed for polishing. *Bole*, or Lemnian earth, is a red clay, unctuous to the touch, and adheres to the tongue; it was formerly used in medicine, and is now employed as a colour, particularly for earthen utensils. *Terra de Sienna* is a brown clay, employed as a colour by artists and printers. *Lithomarge* occurs in fissures of various rocks.

The most valuable of all clays is the *porcelain earth*, or Kaolin, ($3Al_2O_3$, $4SiO_3$, + 6HO,) which, as will be seen hereafter, consists of disintegrated felspar, and forms large earthy masses, which are white, or only faintly tinted, and perfectly free from iron. This valuable material, which is used in the manufacture of porcelain, is found, though not frequently, in layers in granite and other rocks. Superior kinds are obtained from Cornwall, Schneeberg, Meissen in Saxony, Passau, Carlsbad, Limoges in France, and from many other places. That this earth is found in China and Japan is proved by the importation of the first porcelain from these empires, and also by the name Kaolin having been given to this mineral.

49. Common *clay* is, however, of far more importance to mankind than even porcelain earth. When somewhat similar to the latter it is called porcelain clay; or pipe clay, if it is white; Potters' clay, if coloured and of coarser quality. All clays are unctuous to the touch and adhere to the tongue, since they absorb and retain water with great avidity. They absorb fat and oil still more powerfully, and are hence employed for removing grease spots. Clay is also possessed of a peculiar odour, which arises from its property of absorbing ammonia from the atmosphere. Clay is infusible, and blocks of burnt clay are therefore employed, under the name of fire bricks, for building structures, which are to sustain a high temperature, such as porcelain furnaces, blast furnaces, glass furnaces, &c. Earthy clay is employed for the manufacture of various kinds of pottery (Chemistry, § 88). When clay contains lime in admixture it loses its peculiar properties, particularly its infusibility; it then passes into marl and loam.

In concluding our description of this family, mention must be made of *agalmatolite*, a clay-stone, out of which the Chinese carve their idols (Pagodas), and produce figures which give us anything but a sublime conception of a deity.

6th Family—Felspars.

50. The name *spar* is very old, and was probably chosen to indicate a cleavable crystallised mineral. The minerals of this class bear a great similarity in their composition to the zeolites, if the water contained in the latter be disregarded. Their hardness reaches to 7, their specific gravity to 3·3.

They are mostly possessed of vitreous lustre, are generally coloured, and difficultly fusible before the blowpipe. The most remarkable are as follow:—

(1.) *Felspar* ($KO,SiO_3+Al_2O_3,3SiO_3$), which crystallises in prisms of a great variety of forms, belonging to the oblique rhombic system of crystallisation (fig. 14). Figs. 48 and 49 represent two of its usual crystals. It is easily cleavable, and has an uneven fracture; H. = 6; Sp. Gr. = 2·5: it is transparent, of vitreous lustre, white or flesh-coloured, and occasionally green. It occurs in aggregations of well-defined crystals, as also in large crystalline masses. It is found most frequently as a constituent of various kinds of rocks, particularly of granite, gneiss, and syenite, which renders it of particular importance. A bluish-green felspar, of peculiar internal nacreous lustre, is termed *adularia*, or *moonstone*. The amorphous, compact felspar is called felspar rock or *felsite*. This forms likewise a principal constituent of various rocks.

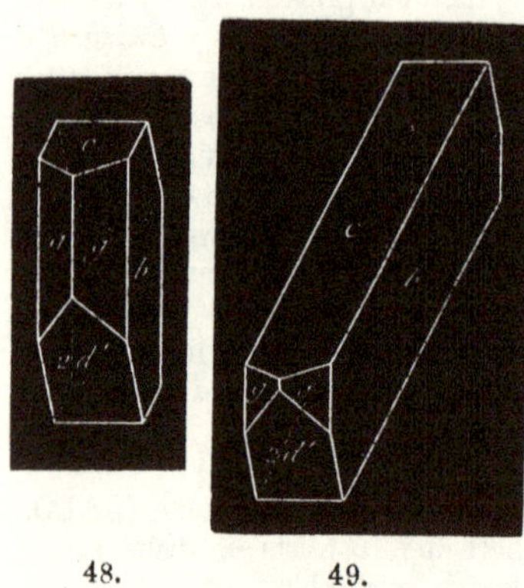

48. 49.

(2.) *Albite* ($NaO,SiO_3+Al_2O_3,3SiO_3$) is felspar, containing soda instead of potassa. It is likewise an important constituent of many rocks. *Spodumene*, or oligoklase, is similar in composition. *Labradorite* is remarkable for its opaline reflections, of a blue, yellow, or red hue, somewhat resembling the colours observed on the breasts of pigeons and on many butterflies. Besides these varieties, we may mention *anorthite*, *leucite*, *nepheline*, *sodalite*, and *hauyne*.

(3.) *Lazulite*, or *Lapis-lazuli*, is distinguished by its magnificent blue colour. It is found in Siberia, Thibet, and China, and is extensively employed in jewellery for ornamental works; and, when properly prepared, is used as a beautiful pigment, under the name of *ultramarine*. Since, however, chemists have become accurately acquainted with the constitution of this mineral, they have succeeded in preparing the above colour artificially. (Chemistry, § 89.)

The following minerals appear to be mixtures of silicic acid and felspar, which have become fused together, by a high temperature, to vitreous, slaggy, or spongy masses. *Obsidian* occurs in black, blue, or greenish-black vitreous masses, and is employed for the manufacture of ornaments, such as boxes, buttons, &c. The South Americans employ this mineral for the manufacture of knives, weapons, &c. *Pumice-stone*, which is found in stream-like layers in the vicinity of volcanos, is very porous, fibrous, and vitreous, and is employed, as is well known, for cutting and polishing, particularly softer objects, since its hardness is only = 4·5. *Pearlstone* and *pitchstone* likewise belong to this family.

7th Family—Garnets.

51. This family embraces minerals of very remarkable crystalline forms; they do not, however, occur in large quantities, and are not applicable to the arts, excepting to jewellery. Their Hardness varies from 5 to 7·5; their Sp.

Gr. from 2·6 to 4·3. They are mostly coloured and fusible before the blowpipe. Besides *Wernerite* and *Axinite*, the latter remarkable for the peculiar form of its crystals, fig. 17, *Tourmaline*, or schorl, is particularly worthy of mention. The latter crystallises in very complicated forms, which are derived from an obtuse rhombohedron of the hexagonal system. The usual forms are six-sided prisms, so much distorted as to resemble three-sided prisms, generally perfect only at one end. Its chemical composition cannot be well expressed by a formula; it contains many constituents, among which are boracic acid, alumina, and silicic acid. It is worthy of mention that a crystal of tourmaline, when warmed, becomes negatively electric at one extremity and positively electric at the other. Tourmalines are found of all colours: the transparent green and brown crystals are employed in the investigation of certain phenomena of light.—It may be remarked of *Staurolite* that its crystals frequently occur in regularly-formed crosses (fig. 50).—The best known mineral of this group is *Garnet*, which crystallises in beautiful rhombic dodecahedrons (fig. 51), belonging to the regular system. It consists of silicate of alumina, combined with another silicate of a metallic oxide, generally lime or iron, but which varies exceedingly, so that a whole series of different garnets are known, like the alums (§ 45), corresponding pretty accurately in their physical characters; and many of them occurring together in the same mass. Garnets are very imperfectly cleavable, their fracture is conchoidal, H. = 6·5 to 7·5; Sp. Gr. = 3·5 to 4·2; they are mostly transparent, and occur of all colours. The beautiful deep-red garnet (*Precious* Garnet) is the most highly-prized variety, and in great request for necklaces, earrings, &c. The greater number of garnets comes from the neighbourhood of Kulm in Bohemia. *Pyrope*, *Idocrase*, and *Epidote* are other remarkable minerals which belong to this family.

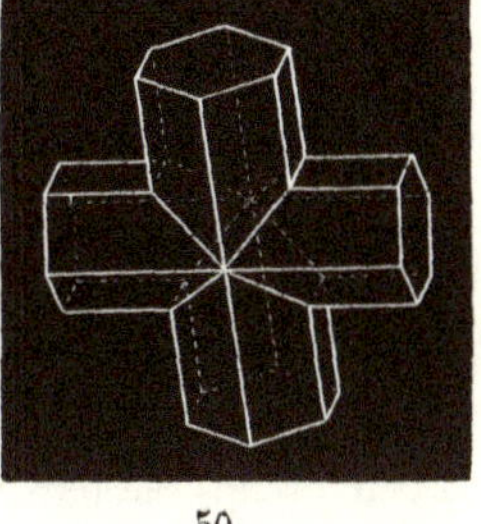

50.

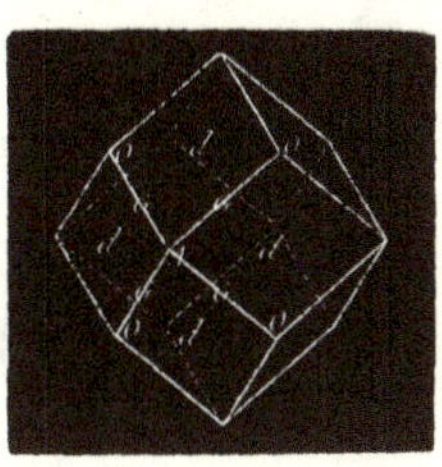

51.

8th Family—Mica.

52. The greater number of the minerals of this family is crystallised in small thin laminæ, of pearly lustre. These laminæ are very cleavable, pliable, and possessed of a low degree of hardness; hence the varieties of mica are smooth and unctuous to the touch. Their chemical composition cannot be expressed by a formula: silica and alumina are the predominant constituents; many of the varieties contain, however, a considerable quantity of magnesia. Mica is often colourless, sometimes coloured, particularly green and black.

Common or *Potassa Mica* is very largely distributed, particularly in various rocks—for instance, in granite, gneiss, and mica-slate, in which it is observable as lustrous laminæ. It occurs in Siberia in very large plates, which are employed instead of glass for windows. Of the various kinds of mica we may mention *chlorite*, remarkable for its fine green colour, which it imparts to those kinds of stones of which it is a constituent, for instance, to *chlorite slate*. *Lepidolite*, or Rose Mica, which contains lithia, belongs to this family.

9th Family—Gems.

53. This class embraces all minerals that possess properties which adapt them to the purposes of the jeweller—hardness, beauty of colour, brilliancy of lustre, rarity, &c. We have already spoken of the diamond, the ruby, and the sapphire. The other minerals of this family have a Hardness of from 7·5 to 8·5, and a Sp. Gr. of from 2·8 to 4·6 : they are transparent, difficultly fusible or infusible, and generally exhibit beautiful colours. Among them may be mentioned *topaz*, which is generally of a fine yellow colour; *pale green chrysoberyl*; *emerald*, remarkable for its beautiful green colour; and *zircon*, of which the hyacinth-coloured variety is most prized, and has received the name of *hyacinth*. The crystals of the two first-named minerals belong to the rhombic system, and those of the emerald to the hexagonal system. Fig. 41. is the usual natural form of the emerald.

SECOND ORDER.—HEAVY METALS.

13th Group—IRON.

54. Iron forms a very important group, both from the great number of ferruginous minerals that exist and from the large masses in which they occur. Their Sp. Gr. ranges to 8·0 ; the greater number is opaque and coloured, and possesses the hardness of quartz. They are attracted by the magnet, and yield, with borax, a dark-red glass in the outer blowpipe flame, and a bottle-green glass in the inner flame. Regarding their application to the production of iron, sufficient details have been given in the section Chemistry (§ 90). The most important minerals of this group are :—

(1.) *Native Iron*, occurring rarely in layers or veins of inconsiderable thickness, or in grains and laminæ. The most remarkable variety is the *meteoric iron*, consisting of masses of native iron which have fallen from the atmosphere, and which weigh from 171 to 3,000, or even 14,000 pounds. Mention may be made here of the *meteoric stones*, which contain, with few exceptions, native iron, besides other earthy constituents, such as augite, hornblende, olivine, &c.

(2.) *Magnetic Iron* ($FeO+Fe_2O_3$), *Oxidulated iron*, crystallises in regular octohedrons (fig. 1, and often in macles, fig. 52). It is remarkable for its magnetic properties: it also occurs in compact masses of considerable magnitude, forming entire mountain strata. It is one of the most highly-prized ores of iron, being used chiefly for the production of steel.

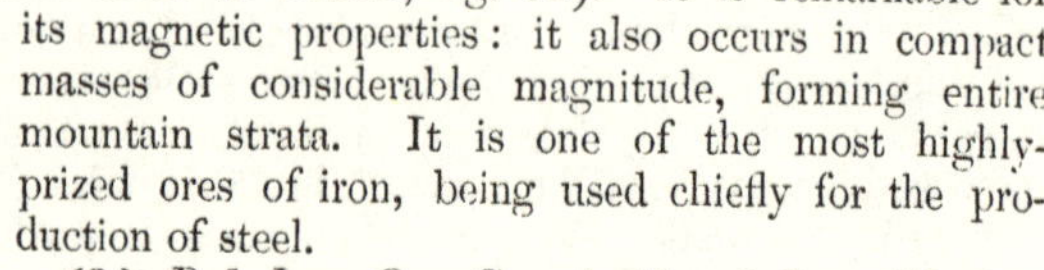

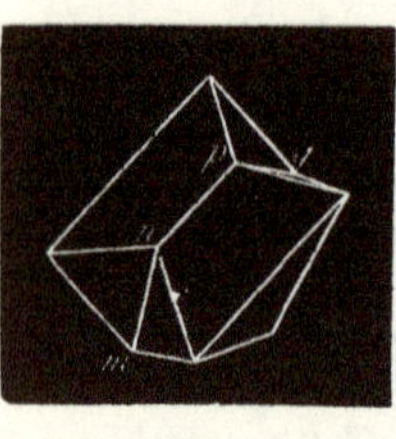

52.

(3.) *Red Iron Ore*, *Sesquioxide of Iron* (Fe_2O_3), or *Red Hæmatite*, crystallises in the hexagonal system as a rhombohedron and its derivatives. It is possessed of bright metallic lustre and red streak, and likewise yields a red powder. It occurs in various forms, as crystallised *iron glance*, micaceous iron, fibrous hæmatite, bloodstone, and also as compact, scaly, and earthy red ironstone, the latter of which is also called *red iron ochre.* If it contains an admixture of clay, it is called in Germany *clay ironstone;* but this is not the important ore that bears that name in Scotland. These minerals are important as iron ores, and are also employed in smaller quantities as polishing materials and as colours.

(4.) *Brown Iron Ore* (Hydrated Sesquioxide of Iron, Fe_2O_3+2HO) does not occur in a distinctly crystalline form. The fibrous brown ironstone, however, consists of fine capillary crystals, radiating from a centre, and forming spherical and botryoidal masses. Besides this variety there is the compact and earthy brown ironstone, which, by containing clay, forms the transition member to the brown and yellow clay ironstones, of which we may mention the yellow ochre and umber, both used as colours. *Pea iron ore,* and *Morass ore,* the ironstone which is deposited in morasses, belong to this class; the latter is less valuable for the amount of iron it contains than the foregoing.

55. Iron occurs combined with sulphur in various proportions, generally as fine crystallised minerals, of a brass-like lustre, which are called *pyrites.* Of these we may mention:—

(5.) *Magnetic Iron Pyrites* (Fe_2S_3+5FeS), which crystallises in six-sided prisms and is attracted by the magnet.

(6.) *Iron Pyrites* (FeS_2), crystallises in the regular system, particularly as a pentagonal dodecahedron (fig. 10) and its modifications; its Hardness is = 6 to 6·5, hence it produces sparks when struck with steel. It occurs very plentifully, and sometimes in very fine laminæ and grains, in coal for instance, and yields protosulphate of iron when oxidised by exposure to the air, particularly in the presence of water (Chemistry, § 93). This salt occurs in the mineral kingdom under the name of *Green Vitriol.*

56. The remaining ferruginous minerals, of which there is still a large number, are most of them of little importance with regard to the quantity in which they occur, and likewise in their applications; we will therefore limit ourselves to a few of the most remarkable:—*Vivianite,* or blue iron ore (phosphate of iron), *green ironstone,* which is the same chemical compound, containing water of hydration, and the series of combinations of arsenic with iron, called *arsenical pyrites,* which possesses a white metallic lustre. Of the latter may be mentioned *arsenical iron, scorodite, pharmacosiderite,* and arsenical iron pyrites, which is also termed *mispickel.*

Carbonate of Iron (FeO,CO_2) occurs in larger quantities; when crystallised it is called *Spathic ironstone.* It forms very obtuse rhombohedrons. This ore is admirably suited for the production of steel. It is also found in the fibrous form, and is then called *sphærosiderite.* The *Clay ironstone* of the Scotch metallurgists consists of carbonate of iron, in combination with variable quantities of carbonate of lime, clay, &c. It is a mineral of great importance.

The *green earth,* which is employed as a colour under the name of *Veronese green,* is silicate of sesquioxide of iron with lime and a little magnesia. *Chrome iron* ($FeO+Cr_2O_3$), which consists of sesquioxide of chromium and protoxide of iron, occurs generally massive, granular, or crystalline, and is important, as being the mineral from which the compounds of chromium are prepared (Chemistry, § 103).

14th Group—MANGANESE.

57. This metal generally occurs as oxide; and, in addition to its being the principal constituent of several minerals, is found in many others in smaller quantities as their colouring matter. The fused minerals are generally coloured violet, whilst the massive minerals are usually brown or black. The most important varieties are:—

Pyrolusite (Binoxide of Manganese, MnO_2), which occurs crystallised in regular rhombic prisms, but is most generally found in crystalline masses consisting of aggregates of acicular crystals. Its colour and streak are black; its Hardness is = 2 to 2·5; its Sp. Gr. = 4·9. The valuable application of this mineral to the preparation of chlorine has already been referred to (Chemistry, § 35).

Hausmannite (Proto-sesquioxide of Manganese, $MnO+Mn_2O_3$), which crystallises in quadratic octohedrons, is brownish black or black, produces a brownish-red streak, and occurs generally associated with pyrolusite.

Braunite, or Protoxide of Manganese, has the same crystalline form as hausmannite; its colour and streak are both dark-brownish black. The value of pyrolusite is naturally much decreased by an admixture of these two minerals; hence, in purchasing this mineral for practical purposes in the arts, particular attention must be paid to the colour and streak.

Manganite (Hydrated Oxide of Manganese) is of little importance in the arts. Sulphide of Manganese, or Prismatic *Manganese Blende*, Silicate of Manganese, Carbonate of Manganese, or Red Manganese, and many other minerals of this family, have not received any application in the arts.

15th Group—COBALT.

58. The minerals of this scarce metal are mostly sulphuretted or arsenical compounds. They are opaque and coloured, and furnish a blue glass with borax before the blowpipe. The most important are: Sulphide of Cobalt (Cobalt Pyrites, Co_2S_3), possessing a white colour, a metallic lustre, and crystallising in regular octohedrons; *Arsenical Cobalt* (Speiscobalt, $CoAs_2$), occurring in cubes, of a white colour, and metallic lustre, in the Erzgebirge in Saxony; *Arsenical Cobalt Pyrites* ($CoAs_3$); *Cobalt Bloom*, or hydrated arseniate of cobalt; *Cobaltine*, or white cobalt ($CoS_2,CoAs_2$), crystallising as pentagonal dodecahedrons, with metallic lustre, and pinkish colour; and, finally, *Earthy Cobalt*, occurring massive or earthy, and of a black colour. The latter contains a mixture of oxide of cobalt, with a considerable quantity of oxides of manganese, copper, and iron. All these minerals are employed for the preparation of cobalt, and especially of the cobalt glass called Smalts (Chemistry, § 95).

16th Group—NICKEL.

59. The minerals of this group are not of more frequent occurrence than those of the preceding group, and they usually occur under similar circumstances. They also generally contain a small admixture of cobalt, sufficient to furnish a blue glass with borax. The most important are:—

Sulphide of Nickel (NiS) which occurs in capillary or acicular crystals; *Red Arsenical Nickel* (Kupfer nickel NiAs), occurring but rarely crystallised, generally massive, dendritic, or botryoidal, and possessing a copper-red metallic lustre; *White Arsenical Nickel* ($NiAs_2$), of tin-white metallic lustre; *Nickel Ochre*, or arseniate of nickel; *Nickel Glance*, or white nickel ore ($NiS_2 + NiAs_2$) of grey metallic lustre. Nickel also occurs in combination with several metals; for instance, it is associated with antimony as antimonial nickel ($NiSb$), antimonial nickel pyrites ($NiS_2 + NiSb_2$), *bismuth nickel pyrites*, and nickel iron pyrites.

All these minerals are but impure chemical compounds, containing always more or less iron, copper, cobalt, lead, &c. Nickel ores are employed for the production of nickel, which is extensively used in the manufacture of German silver. They are found in the Erzgebirge, and also at Riechelsdorf in Hesse.

17TH GROUP—COPPER.

60. This metal forms a large group of minerals, as it occurs not only in great masses, but also in the most manifold combinations. Only a comparatively small number, however, are employed for the production of copper. The Hardness of the minerals of this group ranges from 2 to 4, and their Sp. Gr. to 6; they yield metallic copper before the blowpipe. The following are the most important:—

(1.) *Native Copper*, which seldom exhibits a crystalline form, but generally occurs in peculiar arborescent or moss-like formations. It is frequently found in considerable masses, and is worked for copper.

Red Oxide of Copper (suboxide of copper, Cu_2O) crystallises very beautifully in distinct crystals of many forms of the octohedral system, namely, the cube (fig. 3), the octohedron (fig. 1), the rhombic dodecahedron (fig. 7), the triakisoctohedron (fig. 55), and in many combinations of these forms, as in fig. 53, where the dodecahedron predominates over the octohedron, and fig. 54, where the octohedron predominates over the dodecahedron. Fig. 2 also presents one of the numerous varieties of this mineral.

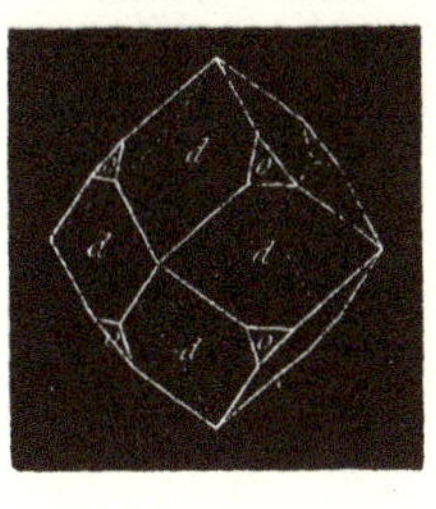

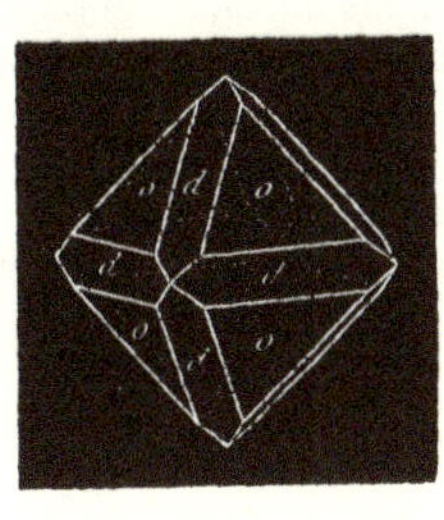

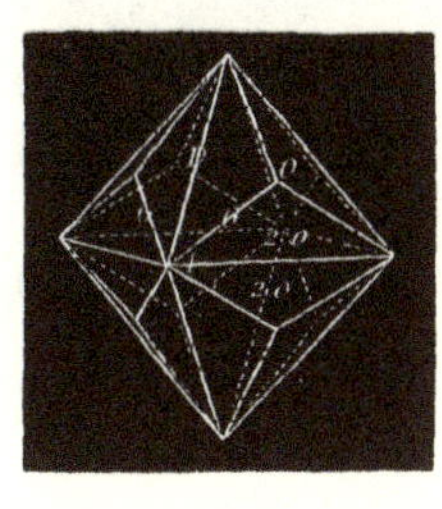

53. 54. 55.

This mineral has a beautiful red colour, but it is generally coated with green. It yields very fine copper.

Vitreous Copper (Sulphide of Copper, CuS) occurs in tabular rhombic prisms of blackish lead-grey metallic lustre, and is worked for copper.

The soluble salts of copper produced in small quantities by the decomposition of other copper ores, particularly of sulphide of copper, are of little importance. They are found principally in the neighbourhood of volcanos, from the fissures of which vapours issue containing hydrochloric and sulphurous acids. Of these salts may be mentioned: sulphate of copper (blue vitriol CuO,SO_3), various phosphates and arseniates of copper, chloride of copper, &c.

The two following may be classed among the most beautiful productions of the mineral kingdom:—

(1.) *Malachite*, or Carbonate of Copper ($CuO,CO_2 + HO$), which crystallises in oblique rhombic prisms, generally uniting into fibrous radiating groups, possesses a fine emerald-green colour, and silky lustre. It also occurs massive and earthy, and is employed for ornamental purposes, and as a pigment; and where it occurs in larger quantities, as in Australia, it is worked for copper.

(2.) *Blue Carbonate of Copper* (azure copper ore) is carbonate of copper combined with hydrated oxide of copper, and occurs either in short prismatic or tabular crystals, or massive and earthy. This mineral is remarkable for its beautiful blue colour, and is hence employed as a pigment. The *Silicate of Copper* ($3CuO,2SiO_3$), Chrysocolla, has a fine green colour.

The minerals in which copper exists in combination with other metals, and in which sulphur is usually a constituent, form a far more numerous group, of which we may mention Bismuthic Sulphide of Copper (needle ore), Antimonial Sulphide of Copper and Lead (Bournonite), Tin Pyrites and *Purple Copper* (Buntkupfererz). The last is a mixture of sulphides of copper and iron. It crystallises in regular octohedrons, and in the forms represented by figs. 3, 52, 2, 4, and 5. It has the lustre of brass, but generally presents the most beautiful variegations of blue and red. The *Sulphide of Copper* (Copper Pyrites $CuS + FeS$) crystallises in quadratic octohedrons, and bears much similarity to the last-mentioned mineral. It is the most abundant of all the ores of copper, and, like the purple copper, is frequently smelted.

In concluding our enumeration of copper minerals, we may mention *Fahl Ore* (grey copper ore), which crystallises in the regular octohedral system, but usually occurs in very complicated hemihedral combinations, of which the double tetrahedron (fig. 56) is one of the simplest. It possesses a grey metallic lustre: its principal constituents are copper, antimony, sulphur, and arsenic, with variable quantities of iron, zinc, and silver. Hence several varieties of this mineral are produced. They are all worked for copper, and the richer specimens for silver.

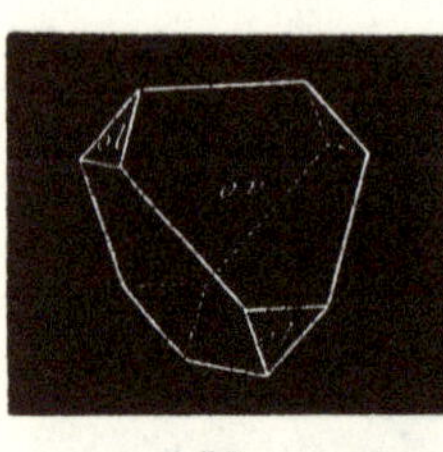

56.

18th Group—BISMUTH.

61. The minerals of this metal are of secondary importance with regard to their distribution and number. Some of the most important are:—*Native Bismuth*, which occurs in regular octohedrons, possessing a reddish silvery lustre; H. = 2 to 2·5; and Sp. Gr. = 9·7; *Bismuth Ochre*, or sesquioxide of bismuth, (Bi_2O_3), occurs in company with the former, particularly in the mountains of Saxony; *Bismuthine*, or sesquisulphide of bismuth (Bi_2S_3) crystallises in rhombic prisms, of a lead-grey metallic lustre. Bismuth Blende consists of silicate of bismuth, and possesses the highest specific gravity of all the ores of this group (5·9). Bismuth has met with but few applications. It is a usual ingredient of fusible alloys.

19TH GROUP—LEAD.

62. This metal rarely occurs in the native state, but generally in combination either with oxygen or sulphur in minerals of low degrees of hardness, but of high specific gravity (4·6 to 8). These combinations yield metallic lead, or the yellow oxide, with great facility before the blowpipe. Many of the minerals of this group occur only in inconsiderable quantities, such as *native lead*, *minium*, or lead ochre, *chloride of lead*, and many others.

On the other hand the *Sulphide of Lead*, or *Galena* (PbS), is the most abundant mineral of this group, and is the one which is principally worked for lead. With the applications of this metal we have already become acquainted. Galena crystallises in the regular system, particularly in cubes, octohedrons, and triakisoctohedrons, and the various modifications of these forms; it likewise occurs in compact masses, which are more or less finely granulated or dense. This mineral is always distinguished by its high specific gravity (reaching to 67·), its grey colour, brilliant metallic lustre, and easy cubical cleavage.

Galena frequently contains silver, in sufficient quantity to render it worth extracting (Chem. § 107). It is likewise occasionally found to contain gold, antimony, iron, and arsenic.

An extensive series of minerals is formed by the combination of lead, antimony, and sulphur, in various proportions. Of these we may mention *Zinkenite*, *Jamesonite*, Sulphide of Antimony and Lead, &c., most of which are named after the discoverers.

Of the native Salts of Lead we may mention sulphate of lead (PbO,SO_3), which crystallises in rhombic prisms, and is distinguished by its brilliant lustre and white colour; White Lead ore, or Carbonate of Lead, which crystallises in regular rhombic prisms, and is remarkable for its adamantine lustre and double refractive power. We shall pass over the combinations of lead with the rarer elements, merely mentioning Chromate of Lead (Chem. § 103), which occurs in a beautiful crystalline form in the Uralian mountains.

20TH GROUP—TIN.

63. Tin does not occur native, but generally as *Tinstone*, which is the binoxide of this metal (SnO_2). This mineral crystallises in quadratic octohedrons, the modifications of which are frequently found in twin crystals. They vary from semi-transparency to opacity, possess a high lustre, are sometimes white, but more generally coloured, and sometimes even black. *Fibrous Tin Ore*, which likewise consists of binoxide of tin, occurs in much larger masses, of delicately fibrous structure. Cornwall and the East Indies are particularly rich in tin ores, the metal from which may easily be obtained by fusion with charcoal.

21ST GROUP—ZINC.

64. Oxide of Zinc is occasionally found in the form of crystalline masses of a red colour, and is called *Red Oxide of Zinc*. A much more plentiful mineral of this group is *Zinc Blende*, which consists of zinc and sulphur (ZnS). It crystallises in the regular system, its most usual forms being the rhombic dodecahedron (fig. 7), the cube (fig. 3), the octohedron (fig. 1), the tetrahedrons (figs. 8 and 56), the macle (fig. 52), and the complex form represented by

fig. 57, in which the cube is modified by the planes of the rhombic dodecahedron (fig. 7), and the tetrahedron (fig. 9). The fracture of zinc blende is conchoidal; H. = 3·5 to 4; Sp. Gr. = 4·1; it possesses an adamantine lustre. Its colour is green, yellow, red, brown, or black. It is worked for zinc, and occurs laminated, fibrous, radiated, and massive.

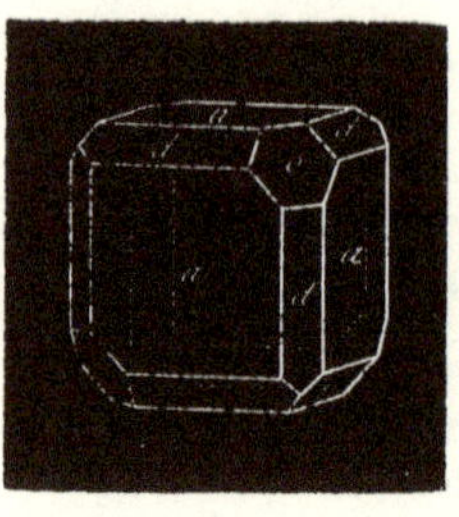

57.

Sulphate of Zinc (ZnO,SO_3) is also found, though only in small quantities, but the Carbonate of Zinc, or *zinc spar*, occurs more frequently. The latter crystallises in the hexagonal system, in the form of rhombohedrons; it possesses a vitreous lustre, and is generally white or only slightly coloured. It is employed chiefly in the manufacture of brass; *Calamine* (silicious oxide of zinc) is the most common mineral of this group, and is employed for the same purpose; it consists of oxide of zinc and silicic acid, and crystallises in rhombic prisms. This mineral is possessed of a remarkable lustre, and is either white or slightly yellow. When heated, the crystals of this mineral exhibit polaric electricity in a remarkable degree, and likewise acquire luminous properties by friction.

22nd Group—CHROMIUM.

65. It is highly remarkable that this metal, of which the chemist prepares a great number of the most beautifully coloured compounds, should only occur in a comparatively small number of natural combinations. This may in some measure explain the circumstance of chromium having been discovered so recently as 1797. In addition to Chromate of Lead (§ 61), and Chrome Iron ore (§ 55), already referred to, we have only to mention *Chrome Ochre* (sesquioxide of chromium, Cr_2O_3), which occurs but rarely and in small quantities. There are, however, several other minerals which contain a small quantity of chromium.

23rd Group—ANTIMONY.

66. The minerals of the antimony-group are possessed of Hardness reaching as high as 6·6; and a Sp. Gr. = 4. Before the blowpipe they yield white vapours, which form a bluish-white incrustation upon charcoal. The rarer minerals of this group are:—*Native Antimony*, *White Antimony* (teroxide of antimony, SbO_3), and *Antimonial Ochre* ($SbO_4 + HO$).

The *Tersulphide of Antimony* (SbS_3) occurs more abundantly, and is a combination of antimony with sulphur, which crystallises in the prismatic system. Its crystals are mostly long, columnar, and acicular, aggregated together, and generally possess a lead-grey metallic lustre. This mineral is employed in the preparation of metallic antimony, and is also used in medicine.

Red Antimony is a compound of oxide with sulphide of antimony, and is distinguished by its cherry-red colour, and the adamantine lustre of its acicular crystals; it is one of the rarer ores of this metal.

24th Group—ARSENIC.

67. This poisonous metal occurs in many metallic compounds, with the greater number of which we have already become acquainted, for example with

Arsenical Iron, Arsenical Cobalt, Arsenical Nickel, &c. The minerals of the arsenic-group yield white fumes before the blowpipe, which have a powerful odour of garlic. The white fumes consist of the highly poisonous arsenious acid. The odour is produced by vaporised metallic arsenic. The most remarkable minerals of this group are:—

Native Arsenic, which is not of unfrequent occurrence; it is generally found in nodular masses not crystallised. It possesses a tin-white or grey-metallic lustre, but soon becomes black by exposure to the air; H. = 3·5; Sp. Gr. = 5·7. It frequently occurs mixed with antimony and silver.

Oxide of Arsenic (AsO_3) may be considered as a product of the preceding mineral, occurring only in inconsiderable quantities, and generally in irregular forms, having an adamantine lustre and whitish colour.

Realgar (AsS_2), is the lower sulphide of arsenic; it crystallises in oblique rhombic prisms, but also occurs in compact masses. It has a pearly lustre, a bright red colour, and gives a yellow streak. It is employed as a colour, and as a constituent of the white fire in pyrotechny. *Orpiment* (AsS_3) is the higher sulphide of arsenic, which is rarely found in the crystallised state, but generally in uniform masses; its lustre is pearly, and its colour bright lemon yellow; it is hence employed as a pigment (Chem. § 45).

25th Group—MERCURY.

68. Although liquid, this metal occurs native, and is found in the form of larger or smaller globules in the cavities and fissures of clay slate, and carboniferous sandstone, as for instance at Moshellandsberg in Rhenish Bavaria. The greater quantity of mercury, however, is obtained from *Native Cinnabar* (HgS), which occurs in botryoidal and compact masses, H. = 2·5; Sp.Gr. = 8. Cinnabar is opaque, and of adamantine lustre; it possesses a carmine colour, and gives a bright scarlet streak. It becomes black on being heated, but reassumes its red colour on cooling. The principal localities in which it is found are Rhenish Bavaria, Almaden in Spain, Idria in Carniola, Mexico, China, and California.

Native Chloride of Mercury, $HgCl$, is a mineral of less frequent occurrence. The mixture of cinnabar, carbon, and earthy matter, occurring in Idria, is called liver ore, or hepatic cinnabar.

26th Group—SILVER.

69. This is one of the more frequent metals, occurring native, as well as in a great variety of minerals, alloyed with other metals, or combined with arsenic and sulphur. Silver ores yield metallic silver when heated before the blowpipe alone, or with carbonate of soda.

Native Silver occurs either in small crystals, of the cubical system, in crystalline groups, or in a great variety of curious forms, sometimes arborescent or like moss, as also in laminæ, irregular masses, and grains. H. = 2·5 to 3; Sp. Gr. = 10·3. It possesses the common properties of silver; it is, however, generally tarnished of a yellowish or brown colour. It is found in most countries; in Germany it occurs with other silver ores, particularly in the Saxon Erzgebirge. The most important ores that are worked for silver are the following:—

Sulphide of Silver, or Vitreous Silver (AgS), crystallises in the cubical system, but occurs more frequently in irregular forms, of a grey or black colour, and metallic lustre. It is also found as an earthy mineral, under the name of *Black Sulphide of Silver.*

Antimonial Silver, containing from 70 to 80 per cent of silver, occurs in modifications of the rhombic prism. It has a silvery or yellow metallic lustre, but is more generally coated with a black tarnish.

Brittle Sulphide of Silver is a combination of the sulphides of silver and antimony, containing about 70 per cent of silver. It occurs in regular rhombic prisms and irregular masses, possessing a metallic lustre and an iron-black colour. The most important silver ore, however, is *Ruby Silver*, which consists of silver, antimony, sulphur, and arsenic. It crystallises in modifications of the rhombohedron, has an adamantine lustre, a colour ranging from iron-black to crimson, and produces a beautiful crimson streak. H. = 2·5 to 3; Sp. Gr. = 5·5 to 5·8. It contains from 58 to 64 per cent of silver.

Sulphide of Silver and Copper contains about 52 per cent of silver, and occurs in blackish-grey crystals of the rhombic system, possessing metallic lustre.

Besides these we may mention the names of several minerals, which occur more rarely, and are therefore of secondary importance. Chloride of silver (hornsilver), bromide of silver, carbonate of silver, bismuthic silver, sternber gite, polybasite, and many others.

27th Group—GOLD.

70. It is indeed highly remarkable, that the more precious the metals the more they appear to be isolated and separated from the other mineral substances of common occurrence, in the same manner as men of a higher order of intellect, seek to stand aloof from those endowed with capacities of a lower cast. Thus gold is generally found native, either crystallised in the several modifications of the regular system, as represented by figures 1 to 9, and 58, 59, and 60, or in the most varied shapes, such as dentritic, capillary, arbores-

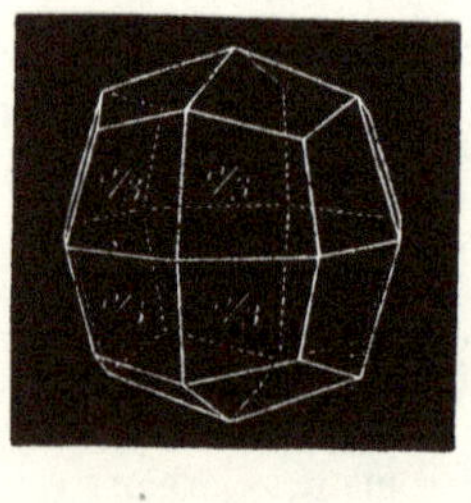

58.

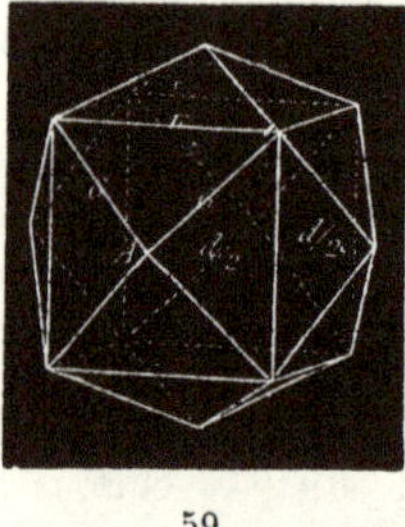

59.

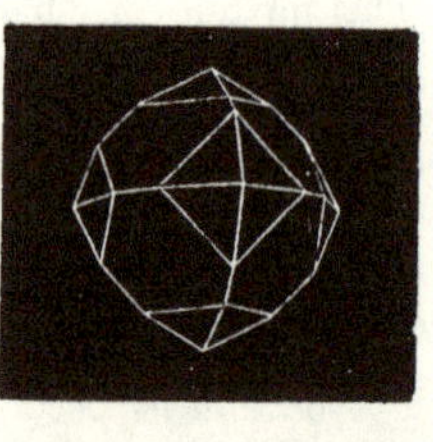

60.

cent, foliaceous, &c. It is likewise found in irregular masses and grains, and finally as sand and dust; it exists in the two latter forms disseminated in various kinds of rocks, such as granite, &c., and owing to their disintegration, it finds its way into the sand of rivers, and the rubble-stones of alluvial soils.

As the specific gravity of gold in this state is as high as 19·4, the smallest grains may be separated from sand by washing, the gold being immediately deposited.

Silver is the metal which occurs most frequently associated with gold: natural alloys of these two metals are found, containing from 0·16 to 38·7 per cent of silver, which causes a considerable difference both of colour and density. In addition to this alloy, we may mention *sylvanite* (graphic tellurium) which contains, besides gold and silver, one of the rarer metals, viz., tellurium.

Europe in general is poor in gold; the only rich gold mines are at Kremnitz, in Hungary. The East Indies, South America, California, and the Ural mountains, are rich in this metal, pieces of gold of considerable size having been found in these localities: in the year 1842 a mass weighing 86 pounds was found in the gold-sand district of Alexandrowsk, near Miask. Pieces of 23 to 24 pounds weight are not unfrequently met with. The most important rivers of Germany, in which gold is found, are the Rhine, the Danube, the Isar, and the Inn.

28th Group—PLATINUM.

71. Platinum is likewise found only in the native state; it generally occurs in nodular pieces and grains, and but rarely in the crystalline form, as cubes. It is frequently alloyed with other metals, more particularly with iron, of which as much as from 5 to 11 per cent is sometimes present. The specific gravity of native platinum is from 17 to 18; its colour is steel-grey. It was first discovered in Spanish America, where it received the name of platina, signifying *similar to silver* (plata being the name of silver). It was afterwards found in quantities in the Ural mountains, where it occurs in alluvial formations, but more frequently in the rubble stones of serpentine rocks. Masses weighing from 10 to 20 pounds have been found in these localities.

THIRD CLASS. MINERALS OF ORGANIC COMPOUNDS.

29th Group—SALTS.

72. As belonging to this small group of minerals we may mention *Humboldtine*, consisting of oxalate of protoxide of iron; and *honeystone* or mellite, a combination of alumina with an acid, consisting of carbon and oxygen (of the formula C_8O_4), which has been named after the mineral mellitic acid. This mineral has received its name from its peculiar honey-yellow colour; it crystallises in transparent, quadratic octohedrons, similar to figures 61 and 62. Both minerals are of rare occurrence, and of little practical importance.

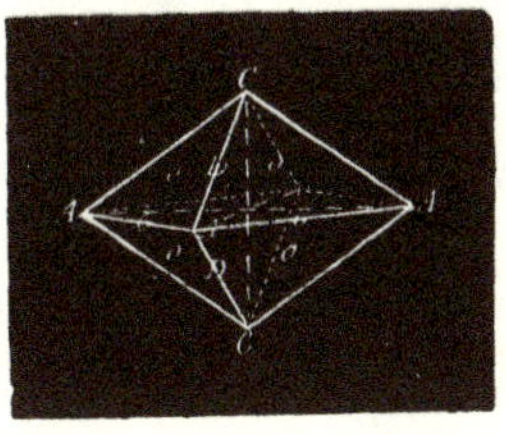

61.

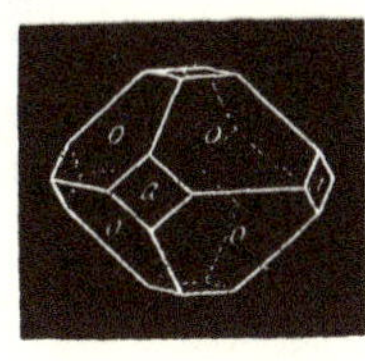

62.

30TH GROUP—EARTHY RESINS. (BITUMENS.)

72*. This group comprises solid and liquid organic compounds, the most important properties of which have been described in the chemical section of this work, among the resins and volatile oils (§ 138). They consist of more or less metamorphosed products of the vegetable remains of a former period, as we have already stated in our chapter on the dry distillation of vegetable matters (Chem. § 169). They occur in the latest formations of the earth's crust. The most remarkable are :—

Amber, a fossil resin, occurring principally in brown-coal formations, and generally in the brown-coal itself. The greater quantity is found in detached pieces on the sea-shore, where it has been washed by the waves, or in the sand and loam, more or less distant from the beach. Amber is fished and dug for more particularly on the east coast of Prussia, from Dantzic to Memel. Pieces of amber are found adhering to fragments of wood and bark: other specimens contain insects, pine-needles and cones enclosed, which leaves no doubt that it originates from a fossilized or an extinct species of pine. Regarding its other properties and applications, see Chemistry, § 141.

Other rarer members of this group are, fossil copal, retinite, mountain or earth wax, elastic bitumen, mountain tallow or Scheererite, idrialite, &c.

Mineral or *Persian Naphtha*, a colourless semi-fluid liquid, is described in Chemistry, § 170, where we have also given a description of *Asphaltum* an *Bitumen*.

[Wollaston's Goniometer, an instrument for measuring the angles of crystals.]

II.—GEOLOGY.

73. In the extensive series of minerals hitherto contemplated, we have not unfrequently met with such as excited our attention, not merely by their individual properties, but by their being dispersed, in considerable masses, through the crust of the earth. Thus quartz, lime, dolomite, and many other minerals, have a limited extent as regularly crystallised objects, but as amorphous formations, they occur in many parts and in immense deposits. Here other more important relations, totally different from those of form, lustre, hardness, or colour, attract our attention. We no longer contemplate those minute and nicely-adapted ornaments of that gigantic structure, the crust of the earth, but have to examine the mighty foundations, walls, and columns, which constitute its fabric.

It is requisite, first, to investigate the materials of this edifice, and after that, its manner of construction.

74. We assume as an established fact that the earth is a spherical body, flattened or depressed at its poles, the diameter from pole to pole being 7,916 miles. The surface of this globe is calculated at 211,000,000 square miles, of which about 150,000,000 are covered with water, and 61,000,000 appear as land. The water, by the law of gravitation, and by the mobility of its parts, assumes a level surface, which appears spherical only when contemplated in its entire mass. If, on the other hand, we examine the solid parts of the earth, it presents itself to the eye in the most varied forms. From the plains stretching out like the ocean, there arise gradually or suddenly considerable heights, sometimes in consolidated masses, sometimes in single chains or isolated peaks,

and these, with steppes, prairies, extensive deserts, table-lands, hills, mountains and mountain vales, abysses and precipitous mural and rocky eminences with craggy summits, lost in the clouds, offer us endless charms by a successive variety of beautiful and sublime scenery.

75. The diversity of the component parts of these mountain masses is, however, scarcely less wondrous than is the remarkable changes of their external forms. A person born and come to maturity amongst volcanic rocks and plutonic formations, accustomed to the daily aspect of granite, basalt, and porphyry, will view with lively surprise the first prospect of the regularly stratified aqueous formations, with their tabular lime and sandstone rocks, containing innumerable petrifactions of organic beings. Hence observation has unceasingly been applied in acquiring a knowledge of the rocky masses. The crust of the earth up to the altitude of 24,000 feet, and down to depths of from 1,700 to 3,000 feet, has been examined in every direction of its accessible parts, particularly within the last 50 years. The indefatigable geologist has successfully applied his hammer to aid his investigations, and everywhere has he collected information, until the science has gradually obtained such a standing as to enable us to form a somewhat definite conception of the structure of the earth, and to account for the co-operating causes of its present form.

Although a more accurate investigation of rocks and their arrangement has, until now, been undertaken only in England, Germany, France, and the adjoining countries, yet sufficient is known of North and South America, and of various parts of Asia, to warrant the assumption, with tolerable certainty, of the following important principle:—*The crust of the earth consists of only a proportionally small number of different rocks, and these are similar to each other at the most distant points of our globe, both as to species and arrangement.*

Thus the various kinds of rock are distributed equally over the entire earth, and the granite blocks of South America, of Heidelberg, and of the most northern latitudes, are exactly alike; while on the other hand, plants and animals of the equator, of the temperate zones, and of the polar circles, exhibit the greatest and most striking differences.

76. Next to this general view of the surface of the earth, a few glances at its interior structure are particularly significant. We have seen that man has penetrated beneath the surface only to a depth most insignificant in proportion to the radius of the globe. Opportunity has nevertheless been afforded by this for making observations, which lead to important inferences. We have noticed in § 127 of Physics, that the average temperature of Germany is from 9° to 10°C. (48° to 50° F.), and that at the equator it is 27° C. (80·6° F.), by which, of course, the temperature at the level of the sea is understood, since the higher elevations always have an inferior temperature. It is a striking fact that a thermometer placed in any locality four feet below the surface of the ground, no longer indicates the change of the daily temperature, but merely that of the year. Again, at a depth of 60 feet it indicates everywhere and at all times the same temperature, which is never affected by the hottest summer or the coldest winter.

Hence this constantly-equal temperature is held to be the specific heat of the earth, independent of that imparted by the sun. Proceeding from this point still deeper, about 120 feet for example, the centigrade thermometer (Physics, § 98) will rise one degree. This remarkable increase in the temperature of the earth towards its centre, amounting to one degree for every 120 feet, has been proved to be the same at the most varying points of the globe, and at all depths.

Now, if the increase of heat progresses in the same ratio towards the deeper and unknown parts, it must attain, at the depth of 36 miles, 1,800° (3,272° F.), a temperature at which iron would melt; at 54 miles a heat of 2,700° (4,892° F.) would prevail, in which all known substances would become molten liquids. Hence it seems but natural to conclude that the interior of the earth is one burning mass of liquid fire surrounded by a crust, that has cooled down gradually and become hardened. We shall see, in the following pages, that there are many other reasons for such a conclusion; we may merely allude here to thermal springs, the waters of which are the hotter the deeper their source may be.

77. A diligent and attentive investigation of the crust of the earth has been undertaken, especially in Germany, where Werner, Professor of Mining at Freiberg, gave the first impulse to the study. We owe, however, the above-mentioned important discovery of the exact similarity of the various kinds of rock, to the illustrious traveller Alexander v. Humboldt, and to the indefatigable Leopold v. Buch.

78. In order correctly to distinguish any kind of rock, we must of course first consider it mineralogically, *i. e.*, its chemical constituents, its hardness, its density, &c., then we have to regard the form of the rock; and although we have no crystals to contemplate in this case, yet when considered in their entire mass, the rocks present, each in its kind, a very peculiar form. Next to this, the peculiarity of their arrangement and stratification is of great importance; and finally, the numerous animal and vegetable fossils enclosed in many of these rocks, contribute most essentially to characterise and distinguish them. Thus we may arrange the subject in the following divisions:—1. *Description of rocks.* 2. *Structure of rocks.* 3. *Stratification* and *superposition of rocks.* 4. *Organic remains.* These four branches constitute DESCRIPTIVE GEOLOGY. After having elucidated these, we may proceed to the consideration of the structure of the earth's crust, the formation of the various chains of mountains, and their connexion with one another, which constitute what may be termed SYSTEMATIC GEOLOGY.

DESCRIPTIVE GEOLOGY.

A. DESCRIPTION OF ROCKS.

79. In endeavouring to distinguish the different kinds of rock, we meet with the same difficulty as in the study of minerals (§ 24). Here, likewise, ocular examination, collecting of specimens, observing the deportment of the rock under the hammer, attentive consideration of the mountains, vales, water-courses, quarries, mines, &c., are absolutely necessary to form a correct conception of the entire subject.

The following description of rocks may, therefore, more correctly be called a mere outline or sketch of the more important members. A collection of the various kinds of rock is much easier to make than one of minerals, as the former generally occur in great masses, and may therefore be obtained at a less cost.

80. The minerals which form a considerable part of the earth's crust are

termed, in general, *rocks*. These rocks, in their internal structure, are of two kinds: either they consist purely of minute particles, for instance, of crystals, grains, laminæ, &c., of one and the same mineral, or they consist of two, three, or four different minerals mixed with each other. There are accordingly two principal classes or divisions, viz., *simple* rocks, and *compound* rocks. Thus, for instance, marble, consisting of nothing but grains of carbonate of lime, is a simple rock. Granite, on the contrary, in which we find quartz, mica, and felspar, is a compound rock.

81. Many terms that have become habitual to us in the description of minerals will have to be repeated also in that of rocks. Granular, spathous, fibrous, foliated, compact, earthy, &c., are terms that have already been frequently used. There are, however, several peculiarities observable in the construction of compound rocks, which we must first notice, before we proceed to describe them. The component parts are combined either in the *crystalline* form, or they are held together in an amorphous state by a non-crystalline mass, in the same manner, for instance, as mortar combines the stones of a wall. In many the cohesion is very great, in others but slight, and these latter are called *loose* rocks, as, for instance, rubble-stones, gravel, marl, &c. The mixture is either distinct and discernible by the naked eye, or it is indistinct, and can be detected only by the help of glasses or by chemical means. A rock is called *slaty* when it splits easily in one direction, which is commonly the case whenever one of the component parts, or all of them, have the form of small laminæ arranged in parallel layers. The structure of the *porphyry* class is very peculiar. This comprises rocks of a given substance enclosing crystals of other minerals, which impart to it a spotted appearance. If a rock contains vesicular cavities filled partly or entirely with another mineral, similar in shape to an almond, it is called *amygdaloidal;* if, however, these cavities occur frequently in it, and are empty, the rock is called *slaggy*. *Geodes*, or *drusic cavities*, are hollow nodules, enclosed in the larger masses of rocks, which are lined inside with beautiful crystallisations. Finally we must mention the *accidental* constituents of rocks, in which occasionally single crystals may be observed; these, however, occur in inconsiderable quantities, and do not in the least alter the specific nature of the entire mass. Thus, for instance, in granite single garnets are sometimes found, the presence of which, however, does not at all affect the character of this species of rock.

Classification of Rocks.

82. Rocks may be classified in various ways; for instance, into granular, spathous, foliated rocks, &c.: it is, however, highly essential that such an arrangement does not separate those rocks that are chemically allied to each other.

The character of a rock is frequently more uncertain than that of a mineral, particularly as one kind or species frequently makes a transition into another; thus, for example, compact limestone passes into granular limestone, and granite into gneiss.

In the following description we shall retain the general division mentioned in § 80, of simple and mixed or compound rocks, and will merely enumerate the most important kinds, with a description of their most striking characteristics.

I. SIMPLE OR UNIFORM ROCKS.

83. These have already been described in the first part of Mineralogy. We will therefore merely recite here the names of those which are most important.

1. *Rock-salt*, § 34.
2. *Gypsum*, § 36.
3. *Limestone*, § 37.
4. *Dolomite*, § 40.
5. *Spathic Ironstone*, § 55.
6. *Pitchstone*, § 50.
7. *Obsidian*, § 50.
8. *Pearlstone*, § 50.
9. *Felsite*, § 50.
10. *Quartz*, § 31.
11. *Augite Rock*, § 42.
12. *Hornblende Rock*, § 42.
13. *Talc Slate*, § 52.
14. *Chlorite Slate*, § 52.
15. *Serpentine*, § 41.
16. *Brown Ironstone*, § 54.
17. *Red Ironstone*, § 54.
18. *Magnetic Ironstone*, § 54
19. *Graphite*, § 30.
20. *Anthracite*, § 30.
21. *Coal*, § 30.
22. *Brown Coal*, § 30, § 23.
23. *Peat*, § 30.
24. *Asphaltum*, § 72.

II. MIXED OR COMPOUND ROCKS.

a. Crystalline Rocks.

25. Clay-slate.

84. This rock is an indistinct mixture of very minute particles of mica, a little quartz, felspar, and talc, containing sometimes particles of coal, hornblende, or chlorite, and having mostly the appearance of an uniform mass. It is distinctly slaty, and has a fracture varying from splintery to earthy. It occurs of a greenish-grey, bluish-grey, violet, red, brownish-black, and, when decayed, sometimes yellowish-grey colour. When pulverised it is mostly white, but when coal is present it is black. Chiastolite, staurolite, garnet, tourmaline, and iron pyrites, are accidental constituents of this rock.

Varieties: common clay-slate; greywacke-slate; dark-grey slate, which is used for covering roofs, for writing-slates, &c.; whetstone-slate; pencil-slate, which is used for slate-pencils and for drawing. The latter containing a considerable quantity of coal, is sufficiently soft to impart its colour to paper, and is employed as natural black chalk. Alum-slate, containing a considerable quantity of coal, iron pyrites, and alumina, is used for the manufacture of alum.

26. Mica-slate.

85. Mica-slate is a distinct admixture of mica and quartz in alternate layers, the mica frequently enclosing small laminæ of quartz. It occurs slaty, grey, white, yellowish, reddish, brownish, and lustrous. Among the accidental constituents are more particularly—garnet, talc, chlorite, felspar, hornblende, tourmaline, staurolite, iron pyrites, magnetic iron-ore, and graphite. It passes over into gneiss, clay-, talc-, and hornblende-slates. The mica is sometimes replaced in this rock by other minerals. The following kinds of rocks being thus produced: talc- and iron-mica-slate; itacolumite, or flexible sandstone, from the mountain Itacolumi in the Brazils; also tourmaline-slate.

27. Gneiss.

86. This kind of rock has received its name from the language of the miners, without any particular meaning; it is a mixture of quartz, mica, and felspar. The quartz and felspar form granular layers, separated by laminæ of mica; it is slaty, grey, whitish, yellowish, reddish, greenish, &c. It forms transitions

into mica-slate, granite, &c. Accidental constituents are: garnet, tourmaline, epidote, andalusite, iron pyrites, graphite, &c.

Talc-gneiss contains talc in the place of mica.

28. Granite.

87. The granular aspect of this rock acquired for it, at an early date, the above name, which is derived from the Latin *granum* (grain). Granite is a mixture of quartz, felspar, and mica, in which, however, the laminæ of the latter do not lie parallel to each other, thus preventing a slaty structure; it is grey, reddish, yellowish, greenish, and white. Accidental constituents are: tourmaline, hornblende, andalusite, pinite, epidote, garnet, topaz, graphite, magnetic iron ore, tin ore, &c. It forms transitions into gneiss, syenite, and porphyry.

The following are varieties of this rock:—*Porphyritic granite*, containing single large crystals of felspar; *graphic granite*, so called on account of its marks, which bear a resemblance to writing, and which are formed by the close intermixture of the quartz and felspar; *protogine*, a mixture of quartz, felspar, and talc; *granulite*, mostly a slaty compound of felspar and quartz; *greisen*, a mixture of quartz and mica, mostly containing tin ore and arsenical pyrites.

Granite is particularly adapted for constructing roads on account of its hardness; it is less suited for building, being rather difficult to work. The city of Aberdeen is built of granite. It is frequently employed in large blocks for bridges, foundations of buildings, monuments, &c. Disintegrated granite yields a productive soil.

29. Syenite.

88. Syenite is a distinct mixture of felspar and hornblende, frequently associated with quartz and mica; the entire mass might, therefore, come under the denomination of hornblende-granite. An admixture of very minute crystals of titanite is likewise characteristic of this rock; it is granular, reddish, or greenish. Its accidental constituents are the same as those of granite. It forms transitions into granite, hornblende, and porphyry. *Porphyritic* and *slaty syenite* are varieties.

Syenite is applied to the same purposes as granite, to which it is, however, preferred for ornamental architecture, on account of its being more finely marked. The numerous and great architectural monuments in Upper Egypt are constructed of a reddish syenite, from *Syene*, from which locality the name of the rock is derived.

30. Greenstone.

89. This rock, likewise designated as *greenstone-slate* (trap, diabase, whinstone), is either a distinct or indistinct mixture of amphibole (bronzite, hypersthene, schillerspar), with felspar, and is either granular or compact, slaty and porphyritic; sometimes it is vesicular or amygdaloidal, the vesicular cavities being filled with calcareous spar. The colour varies from green to black; sometimes it is dark-grey. The more frequent accidental constituents are: iron pyrites, quartz, mica, garnet, epidote, and magnetic iron ore. The amygdaloidal and other greenstones that occur abundantly on the banks of the Clyde, in Scotland, abound with beautiful minerals belonging to the zeolitic class. The localities of Kilpatrick and Kilmacolm are particularly famous for prehnite,

Thomsonite, cubicite, mesotype, harmotome, stilbite, and other minerals of this class. Its varieties are: *diorite*, a distinct compound of hornblende and albite, frequently with iron pyrites (the same rock of slaty structure is called *diorite slate*); *aphanite*, a compact and apparently uniform mixture, containing amphibole and albite, sometimes amygdaloidal, and when there is a preponderance of separate crystals of albite or hornblende, forming a transition into aphanite-porphyry; *gabbro*, a granular mixture of labrador and diallage, sometimes containing titanic iron and serpentine; *wacke*, a brownish or dirty-greenish rock, from compact to earthy, sometimes vesicular, slaggy, or amygdaloidal, originating most likely in the decomposition of various kinds of greenstone. These species of greenstone are used for building, and some of them, which partly pass over into the porphyry variety, are employed in works of art under the name of *porfido verte antico.* Greenstone, on account of its extreme toughness, offers a valuable material for the formation of macadamised roads.

31. Porphyry.

90. Porphyry is a compact felsite mass, containing single crystals of felspar, quartz, more rarely mica or hornblende, and more accidentally garnets or iron-pyrites. Its structure is porphyritic (comp. § 81); it occurs of a reddish, yellowish, and brownish colour, and variegated. Several works of art constructed by the ancient sculptors in stone, designated by this term, do not agree with what is now termed porphyry.

All kinds of porphyry are much used for building, for roads, &c. By disintegration they generally yield a very productive soil, containing potassa. The different varieties are: *quartz-porphyry*, or red porphyry *(porfido rosso antico)*, consisting of a compact mass of felsite, with crystals of quartz or felspar, and mostly yellow, red, or brown; *mica-porphyry*, a mass of compact felsite, with crystals of mica and felspar; *syenite-porphyry*, a mass of compact or crystalline felsite, with crystals of felspar and hornblende; *pitchstone-porphyry*, fundamentally composed of pitchstone, blended with crystals of vitreous felspar and quartz.

It is worthy of remark that several of the finely-spotted porphyries are employed in the construction of works of art, such as columns, slabs, vases, urns, bowls, &c., not unfrequently of extraordinary size. The most celebrated are the porphyry works of Elfdalen, in Sweden, and of Kolywan, in Asiatic Russia.

32. Melaphyr.

91. This rock may be called augite-porphyry, or black porphyry, and also amygdaloid. It is a compact, or somewhat crystalline, and mostly indistinct mixture of augite and Labrador felspar, frequently porphyritic, with single crystals of Labrador and augite, which impart to it a dark-brownish, greenish, or black colour. Accidental constituents of this rock are mica and iron pyrites, but *never* quartz. We may mention as varieties the compact, *porphyritic melaphyr*, and likewise *amygdaloid.* In the latter the principal and generally uniform mass contains vesicular cavities, partly or wholly filled up. These cavities are either quite irregular in shape, spherical or oblong in one and the same direction, or they are pear-shaped with the tapering extremity undermost. No doubt can be entertained that these cavities originated in an evolution of

gas from the interior of the rock. The contents of the vesicular cavities consist of calcareous spar, calcedony, agate, quartz, zeolites, chabasite, &c. The layers or nodules of these crystals are in some cases parallel to the sides of the cavities; in others in irregular masses. The uniform contents assume either botryoidal or stalactitical forms.

Melaphyr is likewise used for building and for roads. It does not easily decay, but by disintegration it yields a very productive soil.

33. Basalt.

92. This rock is generally an indistinct mixture of augite and felspar; it is also called *basanite*, and some kinds of it have received the name of trap. The above constituents are generally associated with olivine and magnetic iron ore.

Basalt is compact, porphyritic, granular, amygdaloidal, and slaggy; its colour is either black, greenish, greyish, or brownish-black. It is commonly hard and heavy. A distinction is made between the common basalt, which is compact and apparently uniform in mass, and *dolerite*, a distinctly mixed basalt, in which we recognise especially augite and felspar. The accidental constituents are nepheline, leucite, mica, and iron pyrites, besides olivine and magnetic iron ore. The *amygdaloidal basalt* possesses vesicular cavities. Basalt furnishes the best material for paving roads. For building, the compact basalt is too heavy, while on the other hand the porous basalt is well adapted for this purpose; it is not applied to finer works of art. The latter kind is met with in Germany, in the vicinity of extinct volcanos, especially in the seven mountains, of which the Drachenfels is the most celebrated; it is found likewise in the most southern parts of the black forest (Kaiserstuhl), and in Bohemia, where it is used as dry building stone, and the lighter variety in the construction of cupolas and vaults. The porous basalt, from the quarries in the neighbourhood of Coblentz (at Neidermending), is much celebrated, and is employed for millstones. When disintegrated by atmospheric influences, basalt yields a highly productive soil, which is particularly warm on account of its dark colour.

34. Phonolite.

93. This rock is called klingstein, or sounding stone, from its property of producing a clear sound when struck with a hammer, and though apparently uniform, it is a mixture of felsite and natrolite; it occurs compact, laminated, porphyritic, from crystals of felspar, but rarely vesicular. The fracture varies from splintery to conchoidal, and from vitreous to earthy. The colours of this rock are greenish-grey, grey, and blackish-grey. A peculiarity of this kind of rock is, that nearly all its exposed surfaces are coated with a white crust of the disintegrated stone. The accidental constituents are: hornblende, augite, magnetic iron ore, titanite, leucite, and mica. The drusic and vesicular cavities generally contain zeolites. This rock passes over into trachyte, and approaches on the other hand to basalt. As varieties we may distinguish compact phonolite, porphyry slate, and the decomposed phonolite, which is a soft almost earthy rock, and yields a kind of porcelain earth, like the above-mentioned white crust of disintegrated rock. This rock, which frequently splits into plates, is used in building, sometimes even for roofing, and also frequently for paving. The clay soil resulting from its decomposition is but little favourable to agriculture.

35. Trachyte.

94. Trachyte is an indistinct, indefinite, mostly granular mixture, in which felspar predominates. It is nearly always porphyritic, from the presence of vitreous felspar crystals, and generally contains scales of mica and needles of hornblende. It occurs granular, porphyritic, compact, slaggy, and earthy. The fundamental mass is grey, yellowish, reddish, or greenish. For building purposes this rock may easily be dressed with the hammer and other suitable tools; but it decays easily, as has been proved, for instance, in the cathedral at Cologne, the more ancient part of which is constructed of trachyte, from the Siebengebirge. It yields a productive loamy soil for agriculture.

36. Lava.

95. This is an indistinct mixture of augite and felspar, frequently associated with leucite and magnetic iron ore, more rarely with mica, olivine, &c. It occurs granular, compact, porphyritic, and slaggy. Its colour is either dark, brown, grey, reddish, greenish, yellowish, or black. All the glowing masses in general that are emitted in streams from volcanos during eruptions, independent of their composition, are called lavas. The different varieties are—the *basaltic lava*, very similar to basalt, though rougher; *doleritic lava*; *leucite lava; porphyritic lava; slaggy lava;* and lastly the *volcanic scoria*, consisting of detached fragments, and called *lapilli*, or volcanic sand. Lava is particularly distinguished for the remarkably fruitful soil it yields by slow decomposition. This may be a consequence partly of its chemical constitution, partly of its dark colour, and of the evolution of heat and carbonic acid proceeding from the ground near volcanos still in action.

b. Mechanically Mixed Rocks.

I. Distinctly Mixed Rocks.

37. Breccia.

96. Breccia is a combination of angular portions of rocks, enclosed within another mass, which may be termed the uniting medium, or cement. These breccias receive different names, according to their enclosed fragments, or uniting medium. Thus we distinguish, *e. g.* granite-, porphyry-, limestone-, and bone-breccias. From the supposition that some breccias have arisen through the forcible trituration of a liquid mass against a solid one, they are called *trituration-breccias*, as, for example, a mass of porphyry with fragments of clay-slate.

When the uniting medium of the breccia is sufficiently hard, it may be used for building material. A few breccias, which, from the admixture of variegated and differently formed fragments of rocks, present a very beautiful appearance, particularly when polished, are applied in ornamental architecture, and receive different denominations, answering to their appearance. Thus a breccia, consisting of granite, porphyry, and diorite, is called *breccia verde d'egitto*, and the various marble breccias are named *violetta antica*, *dorata*, *pavonazza*, &c.

38. Conglomerates.

97. *Conglomerate* is distinguished from breccia by the *rounded* rocky fragments which are cemented together by an uniform mass. They have received various names, according to their constituent fragments: gneiss-conglomerate,

basalt-conglomerate, *greywacke*, *nagelfluh*, &c. Conglomerates may be used for building and road making. They, as well as the breccias, yield on disintegration a soil, the productive quality of which must, of course, depend upon the nature of the rocks composing them. Thus greywacke-conglomerate yields a strong, and therefore loose, clayey soil. The red conglomerate has a sandy or clayey-combining medium, containing layers of porphyry, gneiss, granite, mica-slate, clay-slate, &c., which remain undecayed in the clayey and sandy soil. Basalt-conglomerate generally yields a very fertile loam- and clay-soil.

39. Sandstone.

98. This rock, so universally distributed and so well known, is a combination of minute and mostly spherical particles, held together by a uniting medium which is scarcely to be distinguished. The particles are principally quartz, and the cement is generally clay, marl, or oxide of iron, more rarely hornstone. We distinguish accordingly *clayey*, *calcareous*, *marly*, *ferruginous*, and *silicious sandstone*.

We call it *conglomerate* sandstone, if it contains isolated and large rubble-stones. Besides grains of quartz, it sometimes contains scales of mica or grains of felspar, hornblende, or green earth. The latter imparts a green colour to it, and hence the name *green sandstone*. There are various other admixtures in sandstone, of which we will merely mention the globular concretions of clay, which are termed *clay-galls*.

Many other names given to sandstone, such as keuper sandstone, lias, &c., refer to the systems of stratification, which we shall describe further on.

In sandstone we possess one of the most valuable materials for manifold uses; it is particularly adapted for building, being very workable. Sandstones of finer grain and uniform colour throughout offer an excellent material for sculpture, and have been employed particularly in the rich and magnificent ornaments of our ancient cathedrals. The colour of sandstone ranges from white through yellow, greenish-yellow, to brownish and brown,—the latter variety being found of great beauty, particularly in Würtemberg. Besides these, red sandstone is also frequently found.

Sandstone is but an indifferent material for roadmaking; but the hardest kinds are used for millstones, grinding stones, and many in the form of flags are used for roofing and paving.

The soil it produces by decay is one of the most unproductive, since it is totally destitute of potassa and soda, and incapable of retaining moisture. Sandstone, in which clay and marl preponderate as a cement, is, of course, more favourable to agriculture.

40. Debris. Gravel. Sand. Crumbled Rocks (Gruss).

99. The term *debris* is applied to a loose accumulation of rocky fragments, like breccias, without cement, whilst by *gravel* or *rubble stones* we understand a collection of rounded fragments of rocks, that may be regarded as conglomerate which is not united by a binding material. *Sand* is a loose accumulation of grains of minerals, mostly of quartz. *Gruss* signifies the loose, unconnected constituents of any compound rock, *e. g.* granite-gruss consists of an uncombined mixture of grains of quartz, mica, and felspar.

II. *Indistinctly Mixed Rocks.*

41. Marls.

100. Although apparently uniform, marls are an amorphous mixture of carbonate of lime and of clay occurring of all densities, from compact to earthy, also slaty, but rarely of a fine grain; the colour of marls is either grey or yellowish, reddish, greenish, bluish, black, white, or variegated. They crumble to pieces in the air, generally very rapidly, and effervesce feebly with diluted hydrochloric acid. According to the preponderance of one or the other constituents, or the admixture of other minerals, we distinguish *common marl*, *calcareous marl*, *clayey marl*, *silicious marl*, *sandy marl*, and *bituminous marl*, which is mixed with bitumen (asphalt), and frequently occurs slaty. Finally, we meet with *cupriferous slate*, a bituminous marly slate of black or dark-grey colour, which is famous for its abundance in those copper ores, mentioned in § 59, and which contains besides cobalt-, nickel-, and silver-ores.

Marl is totally unfit for building purposes, in consequence of its rapid disintegration; it is, however, on this account the more valuable in agriculture. Marl soil is considered the most fruitful, although it must be observed that it should not contain under 10 nor above 60 per cent of carbonate of lime. Poor land and calcareous soils are improved by a dressing of marl. The marl containing a larger proportion of lime, is also burned and used as hydraulic lime or cement (comp. Chemistry, § 81). Marls are particularly found in districts of the more recent formations, *e. g.* in Suabia.

42. Clay.

101. *Clay*, though apparently uniform, is a mixture of alumina with a little lime and silica (comp. Chemistry, § 87). It occurs compact, earthy, soft, and friable; it softens in water and is exceedingly plastic. It is found of all colours, and sometimes even black, owing to the presence of bitumen. We distinguish besides pale common clay, yellow loam, and a loose earthy mixture of clay, lime, and sand (löss), of yellowish-grey colour, distributed more particularly throughout the valleys of the Rhine. *Saline clay* is mixed with rock-salt, and has a dark colour, which is due to the presence of carbon.

Only the clay of the more ancient formations, hardened into stone, is used as a building material. Respecting the use of plastic clay, we have given comprehensive information in § 88 of Chemistry.

43. Fullers' Earth.

102. This term denotes a soft friable mass, probably derived from the decomposition of greenstone; it possesses an uneven fracture, from coarse to fine earthy, and is unctuous to the touch. Its colour varies from grey, greenish, yellow, to white. It forms with water a thin unmouldable paste, which is used by manufacturers for extracting the greasy matter from woollen fabrics. It contains about 10 per cent of clay and up to 60 per cent of lime, and is closely related to the boles.

44. Tufa.

103. Under this name several kinds of rock, not accurately defined, are comprehended, containing rather loose and partly earthy combinations of clayey, calcareous, and sandy constituents. Their colour is mostly grey or

yellowish; sometimes they also enclose a mixture of debris or fragments of compact rocks. Amongst other tufas we notice the following, viz., *trass*, a volcanic tufa, which, mixed with 1½ to 2½ parts of lime, forms cement which sets under water (Chemistry, § 81), and hence has been applied to many important purposes. The trass from the neighbourhood of Andernach is the most celebrated in Germany. The volcanic tufa of Italy, *pausilipp tufa* and *peperine*, or pepperstone, are partly applicable for building; but they are sometimes much injured by the influence of the weather. In the neighbourhood of Naples there exist antique buildings, grottos, &c., constructed of this rock; its disintegration produces an extremely fruitful soil.

45. Humus.

104. Arable or cultivated soil is the superficial stratum of the earth's crust. It is a soil mineralogically undefined, and the product of vegetable and animal nature upon the soil yielded by disintegration of any kind of rock. The remains of decaying organic substances (comp. Chemistry, § 164) are intimately mixed with the particles of the crumbled rock, and impart to it mostly a darker and sometimes a black colour, and highly fertilising properties. Some localities of the earth, however, are entirely destitute of this mould; for instance, where pure lime- or quartz-rock forms the surface, vegetable life, in consequence of the deficiency of nutrition, is either totally absent, or if present, it is developed so imperfectly that humus or organic matter cannot be formed on such soils.

B. STRUCTURE OF ROCKS.

105. When a mass of any variety of rock is before us, there are two modes of considering it with reference to its form; firstly its configuration and relation to external objects, and secondly its interior structure. We accordingly distinguish the internal and external forms of rocks.

Internal Forms of Rocks.

106. We do not anywhere find masses of rock of any extent equally coherent throughout. Even in the most compact and hard kinds we observe divisions and separations which are formed by chasms and fissures. The origin of the latter may easily be illustrated by a mass of moist clay. As it dries, fissures and cracks appear on the surface, while the interior parts contract, as may frequently be perceived on a large scale in all clay-soils during hot summers. The above rocks must, therefore, have formerly been soft, and, contracting as they hardened, must have split in various directions, thus forming larger or smaller divisions. In the former case the rocks may be called *irregularly massive*, and in the latter split or *fissured* rocks.

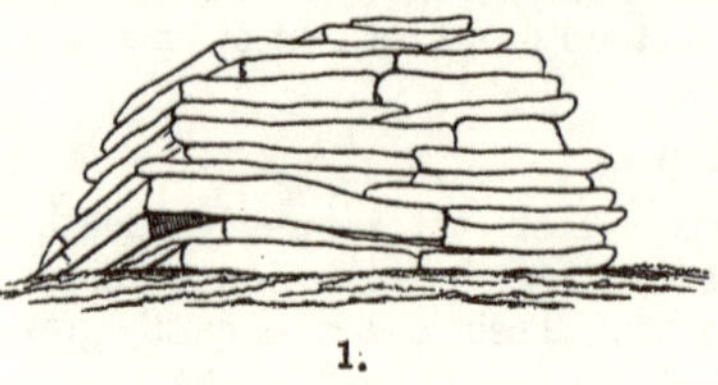

1.

The strata of some rocks are often disposed with a wonderful regularity, presenting a mural appearance as if actually constructed of gigantic masonry. This is exhibited by the Cyclopean granite walls (fig. 1), which occur in prodigious masses on the summit of Goatfell, in the Island of

Arran, where vertical cliffs, appearing like artificial walls, formed of gigantic masses of granite, sometimes rise to the height of 100 feet. There are also rocks enclosing globular concretions, arising from the circumstance that the hardening of the mass has taken place in different central points at once, round which further deposits were formed in concentric layers. Thus when trap-rock is exposed to the weather, the surface undergoes decomposition, and peels off, in thin layers, like the concentric coats of an onion, often exhibiting large round masses on the surface of the rock.

2. Decomposing Surface of Trap-Rock at Corrie, in Arran.

Occasionally the rounded masses are so small, that the rock has the appearance of a quantity of pea-iron ore. More frequently rocks are split up into pillars, having generally the form of hexagonal columns. Such columnar formations are beautifully illustrated by the basaltic columns of Fingal's Cave, in the Isle of Staffa (fig. 3), and by the Giant's Causeway, on the west coast of Antrim, in Ireland. Similar formations occur near Stolpen, in Saxony, and at Unkel, on the Rhine, where columns of the length of from 30 to 80 feet have been observed. These columns frequently have horizontal joints, at which they split into smaller pieces. They are usually accompanied by smaller columns, decreasing in regularity of form, so as sometimes to appear rather as three-sided than as hexagonal columns.

3. Fingal's Cave, Isle of Staffa.

The most common structure of rocks, however, is that of flat or tabular masses. These divisions are, moreover, less regularly defined. They are frequently so massive as to form immense blocks, or they appear as tabular slabs, gradually attenuating to flags and slates.

Stratification.

107. Rocks separated into strata, frequently give peculiar evidence by their structure, that the superincumbent layers did not originate at one and the same period, but that their deposition, solidification, and contraction took place, gradually and successively. This is rendered particularly evident by the circumstance, that between strata of the same kind, intermediate layers are often observed; for instance, beds of limestone are separated by others of marl. We have abundant evidence that rocks stratified in this manner, arose by a gradual settlement of their particles, formerly suspended in water, according to their greater specific gravity. Similar formations of layers or strata may be perceived on a smaller scale on the banks of brooks and rivers. Having in the following pages to return to the origin of stratification, we will first consider some peculiarities of the strata, or of their relative position and direction.

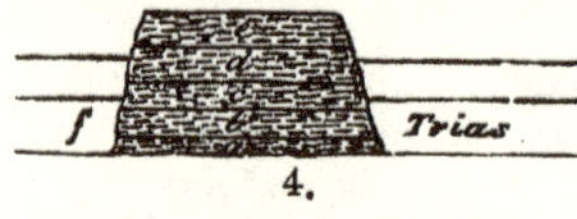

4.

The different layers of a stratified rock, as shown in fig. 4, have a parallel position like the leaves of a book. The thickness or *depth* of the individual strata is exceedingly unequal; for some of them, as represented by lines in the diagram, measure scarcely a quarter of an inch, and are interposed between others that measure in depth from 20 to 30 feet. The direction of these strata is either horizontal, *i. e.* parallel to the surface of the earth, as in fig. 4; or in an inclined position, as the strata *a* to *g* in fig. 5. Various strata are likewise occasionally observed to have a vertical position, as in fig. 6, in which case they are called upright strata. The course in which water poured on an inclined stratified layer would descend, is called, in geological language, the fall or *dip* of the strata. The direction of various strata is designated by the term *strike.*

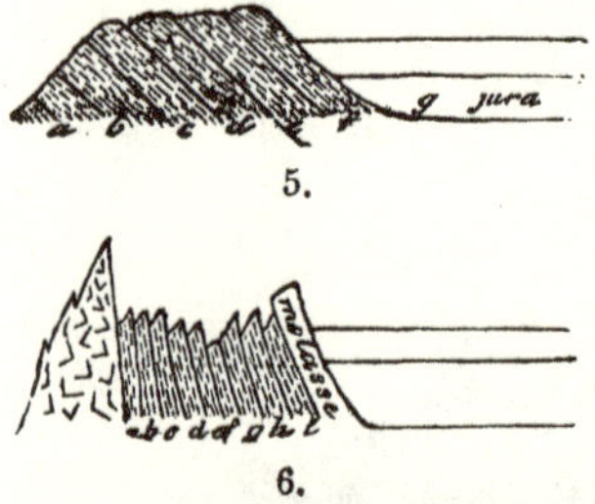

5.

6.

When various strata come to the surface of the earth, as shown in figs. 4, 5, and 6, they are said to *crop out* or *basset.* The exposed or superficial parts of the erect and inclined strata, as in figs. 5 and 6, might be called their *heads.* The horizontal layers become exposed mostly by the action of currents of water, rivers, &c., as in fig. 7, or by the action of the sea; by making railroads, quarrying, and mining.

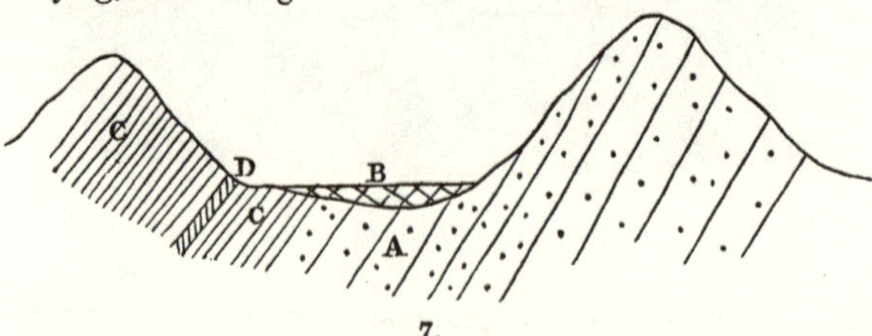

7.

Fig. 7. Denudation of Glen Shirrag, Island of Arran. A, the old red sandstone on the north of Glen Shirrag: B, the alluvial hollow of the Glen: C, the carboniferous formations, with limestone strata: D, apparently unconnected with the old red sandstone.

Various strata are frequently observed to taper off, and to decrease considerably in thickness in one direction, finally either ceasing entirely, or extending further in scarcely recognisable laminæ between the other rocks. This occurs especially with beds of coal, the discovery of the wedge-like end of which frequently leads to a bed of greater thickness.

From this it will be evident that various strata may appear in one place to be in almost immediate contact, whilst at a short distance further they are separated from each other. The erect or inclined strata are evidently not in their original position, but have been forced out of it by some cause acting subsequently to their formation. This is, however, not the only alteration which certain strata have suffered, for their regular and parallel course is often more or less interrupted, and in that case they appear no longer equally superimposed like the leaves of a book, but are bent, twisted, broken up, and intermingled.

External Forms of Rocks.

108. If we contemplate the general aspect of rocks in relation to surrounding objects, they present themselves under three different forms, namely, as *Stratified* rocks, *Massive* or *Unstratified* rocks, and as *Veins*.

Most commonly several strata of different kinds of rock are perceived to overlie each other, and form in this manner systems of stratification, frequently of very considerable magnitude. Limestone, Dolomite, Coal, Sandstone, Clay, and Marl, afford special examples of such stratification.

The structure of the massive rocks never exhibits the slightest appearance of stratification, but merely an irregular division or cleavage, or the fissures mentioned in § 106. They are rarely spread over any considerable space, but occur as isolated precipitous masses breaking through the stratified rocks, and interrupting more or less their regular arrangement. Granite, syenite, porphyry, basalt, &c., exist merely as massive rocks, and never occur stratified.

Veins penetrate not merely through the stratified, but also through massive rocks. Their form may be readily understood, if we study their origin. In the chasms and fissures that arose during the process of hardening of the one kind of rock, there penetrated afterwards the semi-fluid masses of the other kinds, which, in course of time, likewise hardened. These veins are rather irregularly distributed, but their dip and strike are likewise taken into consider-

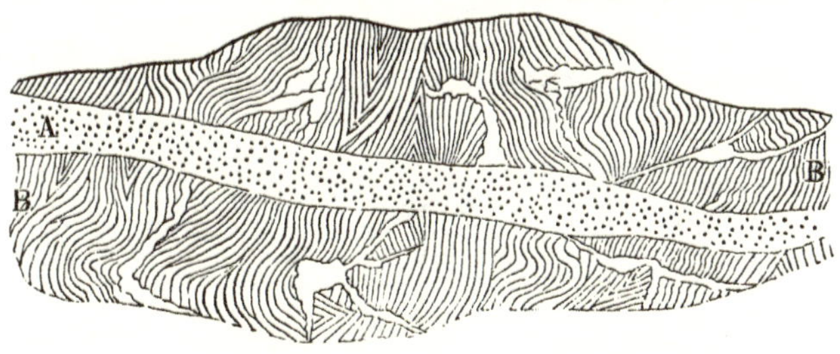

8.

Vein of Granite penetrating contorted strata of Mica-slate near Goatfell, Island of Arran.
A, Granite. B, Slate.

ation. These veins must, however, be distinguished from the mineral or metallic veins, which are generally of inferior thickness, but of more importance, since they contain valuable minerals and ores, and are therefore frequently mined.

Special Forms of Rocks.

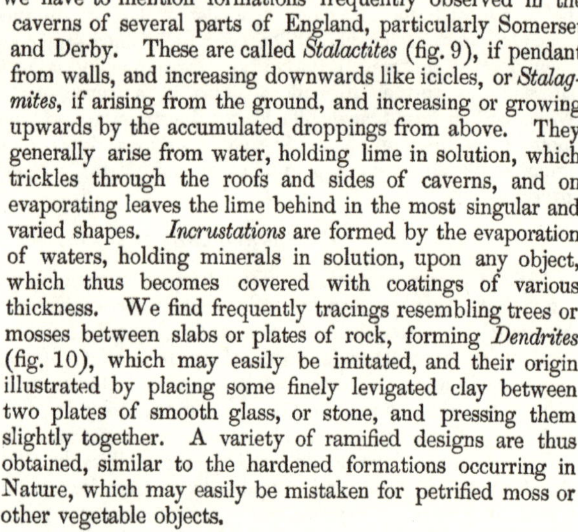

109. As such we have to mention formations frequently observed in the caverns of several parts of England, particularly Somerset and Derby. These are called *Stalactites* (fig. 9), if pendant from walls, and increasing downwards like icicles, or *Stalagmites*, if arising from the ground, and increasing or growing upwards by the accumulated droppings from above. They generally arise from water, holding lime in solution, which trickles through the roofs and sides of caverns, and on evaporating leaves the lime behind in the most singular and varied shapes. *Incrustations* are formed by the evaporation of waters, holding minerals in solution, upon any object, which thus becomes covered with coatings of various thickness. We find frequently tracings resembling trees or mosses between slabs or plates of rock, forming *Dendrites* (fig. 10), which may easily be imitated, and their origin illustrated by placing some finely levigated clay between two plates of smooth glass, or stone, and pressing them slightly together. A variety of ramified designs are thus obtained, similar to the hardened formations occurring in Nature, which may easily be mistaken for petrified moss or other vegetable objects.

9.
Stalactites and Stalagmites.

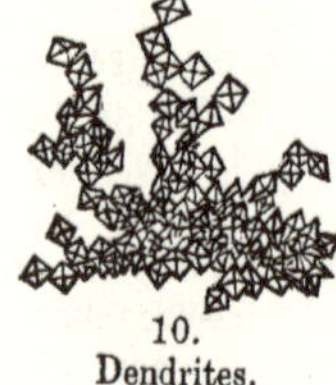

10.
Dendrites.

C. SUPERPOSITION OF ROCKS.

110. From the relative position and combination of the strata, masses, and veins, we are enabled to identify the different periods of their formation, or the ages of the deposits.

The relative position of the strata to each other, may be very different. For instance, they may be lying horizontal and parallel to each other (fig. 11), or they may assume an inclined or vertical position (fig. 12), covered by parallel horizontal stratifications.

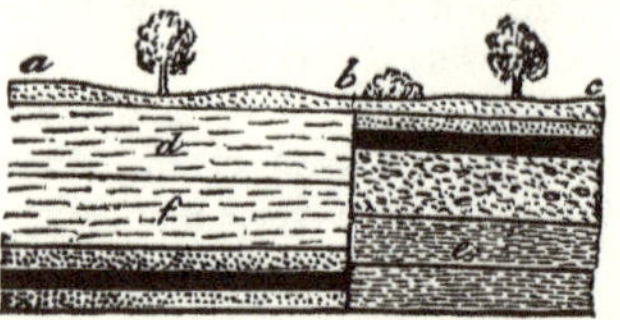

11.

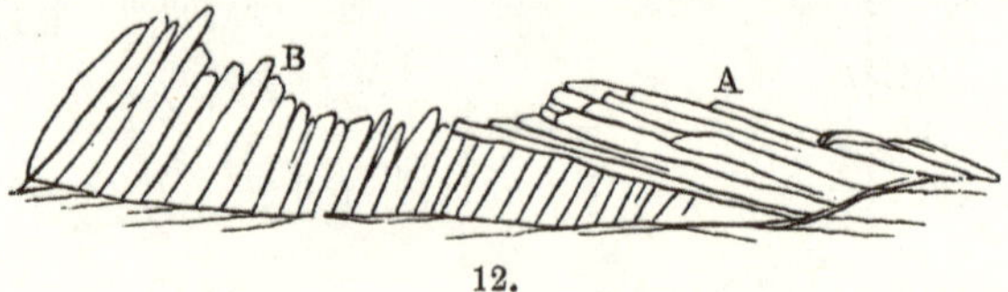

12.

Massive rocks generally are found rising side by side, and it rarely happens that one kind of rock is covered over to any extent, by the horizontal layer of another. Trunk-shaped and *fragmentary* rocks are often partially or entirely surrounded by a layer of another rock; as for example, granite by gneiss (fig. 13), which often occurs when the interior rock, in its eruptive ascent, breaks off, and carries along with it fragments of the other which it entirely encloses.

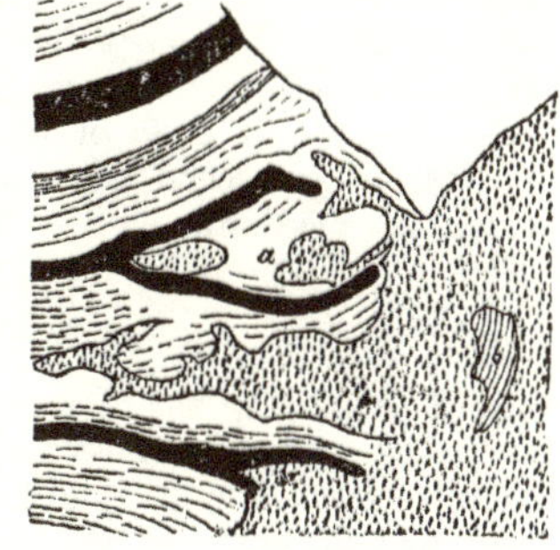

13.

Veins generally extend more in a vertical direction towards the interior of the earth, than in horizontal or oblique directions. They are frequently found to pass through the rock almost perfectly parallel with each other. By a subsequent displacement or disturbance of the position of the main rock, these veins are of course likewise displaced and broken up, which gives rise to great difficulties in mining, in following up a rich vein of ore. The lodes also cross and pass through each other. The coal measures of Great Britain, are frequently seen to have been dislocated by *faults*. This

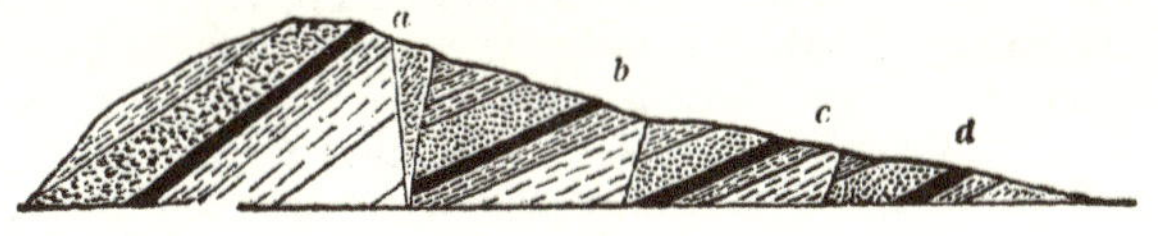

14.

will be easily understood by referring to figs. 11, 14, 15, in which *a*, *b*, *c*, *d* represent coal measures, which have been displaced from their original position, by subterranean disturbing influences.

15.

Fig. 16, represents a very remarkable example from the mountains of Jura, where, owing to the flexibility of the strata, they have suffered great contortion, without becoming ruptured, so as to produce faults.

111. From a closer observation of the above-mentioned relative position of rocks, we gather the most important conclusions, as to which of them is the older, or, what amounts to the same, which formation was hardened first. The following principles may be accordingly accepted as perfectly established. The upper strata are newer, or of more recent date than those which are below them; rocks which have disturbed the position or stratification of the adjacent formations are more

16.

recent than these; rocky masses in the middle of other rocks from which they are separated by sharply defined lines, are generally of more recent formation than the latter; rocks which enclose fragments or disjointed layers are younger than those to which these detached pieces belong; veins and lodes are more recent than the beds of rocks in which they are found, and younger than the veins which they cross or intersect; and, finally, if one stratum of rock is younger than a second and older than a third, the second must be likewise older than the third.

D. ORGANIC REMAINS.

112. Many deposits of rock enclose forms which are called petrifactions, and which may be recognised at a glance, as not being of mineral origin, but to have belonged formerly to the vegetable or animal kingdom. Hence it follows that the origin of the rocks themselves must be dated from the same period in which those plants and animals existed. The petrifaction of these bodies has not been the result of a transformation of their chemical constituents into those of a mineral character, since that would be impossible, as has been shown in Chemistry (§ 10). On the contrary, these plants and animals, when on the surface of the earth, became enveloped, during its great revolutions, in the semi-fluid substance of the rocks in which they are now found, and which subsequently hardened. It is evident that under such violent processes the softer and more perishable parts could not be preserved, and hence, in general, only the more durable parts of plants, such as bark, wood, and ligneous fruits, or the calcareous shells of corals, mussels, and snails, as well as the bones of the higher class of animals, have been preserved. The more perishable organised formations, consisting of carbon, hydrogen, and oxygen, have undoubtedly been sooner or later decomposed, since they are never found in the rock. Nevertheless, under favourable circumstances, many a token or evidence of these formations has been preserved in the midst of destruction. Delicate leaves, and minutely articulated insects enveloped in the semi-fluid mass, have at least left behind impressions in the hardened rocks, from which their formation and class may often be clearly traced. In other specimens the innumerable little interstices or cavities of their bodies have been gradually filled up with the mineral fluid, which, upon hardening, preserved likewise their internal structure.

113. Difficult as it was at first to explain the appearance of an innumerable host of organic remains enclosed in rocks at great depths, and at altitudes of 12,000 feet, these petrifactions became at a later period most essential, as characteristic of the various rocks in which they occur. The following facts have resulted from accurate observations of these remains.

Petrifactions are found only in *stratified* rocks, which have been deposited from water, and never in unstratified or igneous rocks. The number of species, both of petrified plants and animals in the various strata, is very unequal: those occurring in the upper strata, approximate more closely to the still-existing species of the animal and vegetable kingdoms: these, however, decrease in the lower strata, so that the more perfectly developed animals and plants gradually disappear whilst the lower orders prevail, and the existing species become more and more rare. In the lowest and oldest strata, only such fossil remains, of organised beings, are met with, as are now no longer to be found in the recent state.

If the formation of two beds of rock, in different localities, has for other reasons been recognised as contemporary, they must contain the same petrifactions; and, on the other hand, we conclude from the exact similarity of the species of fossils existing in these rocks that they must be of coeval formation. Hence petrifactions have become of the utmost importance in ascertaining the age of the several strata, and in some cases they are the easiest, and even the only means of deciding that question: as we find in the various strata a widely-differing vegetable and animal world, we conclude that the climate and condition of the surface of the earth, at the various periods of its formation, must have been very dissimilar. Again, the fossils of the oldest strata give evidence of the animal creation having been much more equally spread over the surface of the earth than it is at present; and hence the great difference of temperature at the poles and the equator, seems not to have been so remarkable formerly, as it is at the present period.

114. The total number of fossil plants and animals is exceedingly great, and has become the object of two special sciences, namely, *Fossil Botany* and *Palæontology*. Their correct description requires of course a comprehensive knowledge of botany and zoology, and therefore, in treating of these sciences, we have paid proper regard to these petrifactions. However, we will introduce here a concise review of the plants and animals which occur as fossils, beginning with the lower or more imperfect orders.

Of *fossil plants*, we find the following orders: algæ; lichens; and mosses; *Equesetaceæ* occupying the oldest up to the mediæval strata. Lycopodiaceæ, tree ferns, particularly abundant only in the old strata; Liliacea; palms, stems, fruits and foliage; *pines* and *dicotyledonous trees;* the latter occur only in the more recent strata.

Fossil Animals.—*Infusoria* are found in many strata; polypi or *corals* occur most frequently in the oldest formations. Radiata and echinodermata, amongst which are found encrinites, starfish, and the common sea-urchin (*Echinus esculentus*), and *mollusca;* these are the most frequent of all, and to the geologist the most important. There are found, beginning in the old strata, and most plentifully in the middle strata, not only bivalve shells but also univalve snails, and among the latter especially several important genera now perfectly extinct, as ammonites and belemnites. Annellata or fossils of the worm kind are rare; *crustacea* are likewise not of frequent occurrence. *Insects* occur distinctly only in beds of brown-coal, especially in amber: they are on the whole but rare. *Fishes* are exceedingly numerous, upwards of 800 species having already been recognised in the various strata. *Amphibious animals* are represented by the batrachia or frog tribe, though rare; and the ophidia or snake tribe are replaced in great number by saurians or the lizard tribe, now and then of gigantic size, but at present totally extinct. *Birds* are but seldom found in the older strata; *mammalia* exist only in the uppermost strata. There are, however, several extinct species of gigantic size, including the mammoth, megatherium (page 309), dinotherium, &c. Monkeys are exceedingly rare. Traces of human remains are not contained in any of those strata that have been subjected again at a later period to a general destructive influence. Man, therefore, did not appear on earth until its crust was sufficiently stable, and suffered no longer any general revolution.

SYSTEMATIC GEOLOGY.

Origin and Structure of the Crust of the Earth.

115. This wondrous edifice, inhabited by man, did not receive at once its present form. Let us trace, from the preceding statements founded on experience and facts, the history of its origin and progress.

There was a time when the whole earth must have been a liquid glowing mass rolling its onward course through space. The elements or simple substances which it contains then united with each other only in such combinations as could exist at a high temperature. The gases formed the atmosphere which surrounded the firmer nucleus as a covering, with this was associated the vapours of an immense number of volatile compounds which could not remain in a solid or liquid state at such a temperature. The ocean was then in the form of vapour. Thus the earth in its first phases of formation appears to have been a soft, red-hot nucleus, enveloped by an immense and very dense atmosphere which surrounded it or followed its course, perhaps in the manner that the vapoury sphere or tail of the comets and nebulous stars appear now to accompany these bodies through the illimitable universe.

But by continually radiating heat into infinite space, the earth suffered a decrease of temperature at least on its surface. The difficultly fusible chemical compounds, such as silicate of alumina and magnesian clay-slate (mica-slate), &c., began gradually to separate in the form of finely laminated crystals, and by continued cooling to settle upon the surface of the nucleus of the earth, forming the first thin coating or crust over the red-hot liquid mass, and thus separating it from its vapoury atmosphere. This was the commencement of the earth's crust, which might now be increased in firmness more rapidly since the immediate influence of the internal heat was arrested, and as the combinations, existing in form of vapour, might now be deposited thereon, at least partly, in the form of liquids.

116. At that time organic life could not exist. The crust was still too hot to admit of plants taking root and growing; the existence of vegetation, however, is indispensable to animal life, and indeed those lower slaty strata, consisting of mica-slate and clay-state, contain nowhere the least trace of animal or vegetable matter. If water had gathered already at that period upon the crust of the earth, it must have possessed a much higher temperature than at present; hence it was capable of dissolving numerous chemical compounds; and while the ocean at present contains only the easily soluble common salt, &c., the ocean of that period may have held in solution great quantities of silicates, sulphates, and carbonates. It also broke up again a portion of the solid crust, and formed therewith a muddy liquid, which as the earth cooled, again gradually deposited its solid parts in granular strata, forming what is now known as sandstone.

117. Thus we behold acting continually, in concert and by turns, the laws of chemical affinity and of gravitation, in obedience to the latter of which the more compact substances endeavoured to occupy the lowest place. Had this mode of formation thus regularly continued, the surface of the earth must have assumed a tolerably symmetrical shape; the eye would have beheld neither elevations nor depressions; the main body of the earth would have been covered all round by a shallow ocean, and this in its turn would have been enveloped by the atmosphere. The surface of the earth, however, is differently formed. Repeated disturbances gave to it a more varying exterior. And what may have been the cause? The very same powers of Nature, which, by the same laws, prevail up to the present day, and which, acting under the peculiar circumstances, existing at that period on a grander scale, produced phenomena now scarcely conceivable.

118. The more compact parts that were deposited first are justly called *fundamental* or *primitive* rocks; what was formed next in strata, is designated as *stratified* formations, consisting generally of several different strata, which form together a stratified system. Whatever rocks originated within the same period we call *coeval formations*, and hence we speak of the oldest, the mediæval, and the modern formation, which follow each other in consecutive order.

The crust of the earth, upon hardening and contracting, split into fissures and chasms similar to what we perceive frequently on a considerable scale in parched clay soils. The water entered these chasms, widening them more and more by its solvent power, and penetrated at last through the thin crust to the still glowing interior mass. The result of the sudden contact of an immense body of water with a red-hot surface, would be the formation of a vast body of steam, which would attain simultaneously an extraordinary expansive force from the high temperature. These vapours pressing in every direction with an irresistible force, raised the crust of the earth, puffing it up here and there in vesicles of immense size; they tore it up finally, with awful force, and from the opened abyss there poured forth the red-hot liquid mass. Convulsively propelled by the vapours thus liberated, it spread over the neighbouring surface or was formed into mountains surrounding the opening of the eruption.

119. Let us cast a glance on the present surface of the earth. How different do we find it from that regular form described in § 117! From the uplifted portion of the earth's crust the waters have flowed to the lower parts. The solids have separated from the liquids; the former appearing as *continents* surrounded by islands, the latter as the *sea*. The main land itself consists partly of stratified rocks, partly of an irregularly shaped mass, which has been forced up from the interior and slowly solidified, and which hence presents the appearance of an irregular mass of unstratified rock. The fissures that arose here and there in both formations were filled up with the softer rocks or ores, and in this manner originated *veins*. (Comp. § 108.)

We have now recognised *water* and *fire* as the two causes of the above forms, and hence we name the latter from the mythological representatives of the former: Neptunic, or water formations, and Plutonic, or volcanic (fire) formations.

120. The mountains of this period of primitive formation were not of considerable altitude, nor the seas of any great depth. The localities which had become dry were gradually covered with plants, and perhaps coeval with these animals were created. Considering the thinness of the earth's crust at that period, both land and water must have possessed a higher temperature than at the present time, and hence only those beings were created as were capable of existing under such conditions. Ferns, polypi (corals), are the essential remains of the first living creations that are found in the oldest strata, then formed.

121. It is uncertain how long after this first revolution the earth's crust remained in the condition then acquired. It may have been hundreds or thou-

sands of years. The thickness of the strata gradually deposited, and the successive generations of animals, the remains of which lie over each other in the later formations, afford only relative indications with respect to this subject.

It is, however, certain that the first revolution was not the only one. Although the crust of the earth increased in thickness by its continual cooling, still the same causes have effected later eruptions, the essential phenomena of which we have already described. The tension and pressure of the vapours must, however, have become much greater from the increased thickness of the crust that confined them, and consequently the now compact strata have been raised to a much greater height, and the quantity of massive rock forced up through the openings has been much greater, and piled up higher than on their first formation.

The massive rocks of the earlier formation must likewise have been frequently pierced by those of the subsequent periods, whereas the reverse of course could not take place. The waters destroyed at the same time a great part of these rocks and deposited them again in strata, while the vegetable and animal world was overwhelmed in the ruins, and here and there buried and petrified. (§ 112.)

122. Thus several revolutions followed each other at increasing intervals of time. For each later one a greater lapse of time was required in proportion to the still increasing thickness of the crust of the earth, before new fissures, penetrating into the interior, could give access to water. The result was, however, all the more powerful, and the displacement of the strata previously formed, as well as the rising masses of Plutonic rocks, were so much the more considerable. It is an ascertained fact that the *highest* mountains of the earth, the Andes, Cordilleras, Alps, &c., are at the same time the most recent, that is to say, the latest which have been upheaved.

123. Each of these struggles of formation was terminated by the closing up of the fissures and chasms in the crust of the earth, partly through the continued cooling of the interior mass, partly by being covered by aqueous deposits on the outer surface. In some places this was effected perfectly, in others less so, and probably in the latter a new eruption was occasioned at a later period.

But even with the termination of the last general upheaval not all the fissures, leading to the interior, were perfectly closed. In isolated localities where these chasms happened to be very wide, or where mighty rocks accidentally presented gaps between their parts, these openings into the interior were preserved and exist up to the present day. They might properly be compared to the shafts of our chimneys which lead from the exterior of a house to a fireplace.

Such openings in the earth are called *volcanos*, fig. 17. Their operations and effects are pretty well known and easily understood from the previous statements. If their shafts were empty we should be able to look down them to the glowing bowels of the earth; but these hollows or craters are covered with cooled and hardened masses of rocks called *lava*, and with other *volcanic* formations.

From time to time the waters, in a manner not very difficult to explain, find access to the interior of these volcanos. The steam suddenly rises, bursts open large fissures, and causes earthquakes, that thus convulse a large extent of country, and generally precede an eruption. For the increasing tension of the steam will at last force the glowing mass upwards, together with its solid

17. View of Barren Island, in the Bay of Bengal.

cover. The repeated rising and falling of the great volumes of steam, their partial escape, and the violent commotion and vibrations of great masses of the earth, are always attended with terrific noise, which may be compared at times to the continued rollings, and occasionally to the single claps of thunder. The red-hot and liquid mass being forced up finally to the mouth of the crater, its cover is immediately burst and thrown up towards the heavens, its fragments and dust being scattered in the air, and carried by the winds, as volcanic

18. Crater of Vesuvius in 1829.

ashes, often to a distance of many miles, fig. 18. The glowing mass or stream of lava overflows unimpeded the margin of the crater, and in its progress down the sides of the mountain destroys irresistibly every thing it meets.

This terrific revolution of Nature possesses, however, at the same time the conditions of its termination. The steam having escaped, the calm in the interior is restored, and the ejected lava-stream flows slower outside the mountain; finally the progress of the stream is interrupted, and the lava begins to harden, while the interior mass sinks down again to its original level. Only steam, sulphurous vapours, &c., still escape from the crater, and hot fountains spring forth in its neighbourhood, indicating that all below is still glowing. A. von Humboldt truly designates volcanos as the safety-valves of the earth's crust.

124. The environs of volcanos are covered with older and more recent streams of lava, which by decomposition yield a most productive soil, and hence a most luxuriant vegetation surrounds the bases of all volcanos. In spite of the dangerous proximity, several villages have been built near Mount Vesuvius within the reach of its destructive activity. Moreover, in the neighbourhood of volcanos minerals are now in daily progress of formation, either crystallising from the glowing mass, or being formed by the decomposing influence of the rising acid vapours upon other rocks. Hence in these localities a large number of minerals is to be found.

In course of time, however, all volcanos seem to become extinct, as is the case already with many. Thus, for instance, the so-called Eifel, between the river Aar and Treves, consists of a group of volcanic elevations. Lake Laacher, near Andernach, is the crater of an extinct volcano, filled with water, the whole surrounding country bearing characteristic evidence of volcanic origin.

The external form of volcanos is very peculiar, and generally *conical*. They are, in fact, gigantic air or steam vesicles which have been upheaved from below, and finally elongated to an apex, at the termination of which the steam and gas broke through. But such a disruption has not taken place in every case. We find a great many conical mountains that never were active volcanos: in these cases the force acting from below was not sufficiently powerful to pierce through the crust; the glowing mass was hardened inside without even reaching the surface. Indeed, we frequently find in the centre of such conical elevations, consisting of stratified rocks, a mass of Plutonic rock, particularly basalt.

125. In Europe there are no active volcanos of importance, with the exception of Mount Vesuvius, Etna, and Stromboli, in Italy, and of those in Iceland, among which Mount Hecla is the most celebrated. The eruptions of the above-named volcanos following each other at continually greater intervals of time, though still formidable to the nearest neighbourhood, do not now extend over any considerable extent of country. History, however, records several instances of terrible volcanic disturbances, which proved destructive to entire districts, and even to whole countries. Thus in the year A.D. 79 the flourishing and rich cities of Herculaneum and Pompeii were buried beneath volcanic ashes. Lisbon was destroyed by an earthquake in the year 1755, and even at more recent dates formidable destruction by earthquakes has taken place in South America. In that part of the world entire groups of volcanos are found still active, from the position of which L. von Buch points out, that they stand on the fissures of former disruptions of the earth's crust, and have interior con-

nection with each other. The most celebrated volcanos of South America are—the Jorullo, which arose in 1758, and the Cotopaxi, of the chain of Andes. The latter volcano, which is 17,662 feet in height, now and then sends forth great masses of mud and quantities of fish, thus proving in a remarkable and convincing manner its internal connection with the waters.

126. Hitherto we have directed attention only to one of the phenomena that appeared during the early revolutions of the earth, namely, its volcanic disturbances.

Let us now return to other phenomena, and consider the development of animal and vegetable life. It is clear that organic growth could proceed in a proportionately larger scale, the longer the periods were that elapsed between the succeeding disturbances. Plants and animals made their appearance not only more plentifully but also in greater variety. Palms and coniferous plants appear in addition to ferns and equiseta, and batrachians and other amphibious animals, in addition to fishes. Intermingled with these the crustacea appeared in immense numbers. Thus the more perfect creatures followed in proper order upon the imperfect, since the existence of the latter formed the indispensable condition of that of the former. A certain change likewise took place with regard to the formation of rocks. The deposition of the insoluble and difficultly-fusible combination of silica and alumina in the primitive rocks was followed by the gradual deposition, amongst the mediæval rocks, of beds of limestone, gypsum, rock-salt, and of coal, the remains of the destroyed vegetable kingdom of earlier ages.

127. Consequently, it is natural that, in penetrating the crust of the earth, we should meet with a series of strata differing in character according to the period at which they were individually formed, and as in all essential points the same phenomena had occurred over the whole surface of the earth, it follows that the coeval formations of its crust must be everywhere equal or similar. Experience has, on the whole, confirmed this inference, though, in some instances, the proof is often difficult and sometimes impossible to obtain. Thus, everywhere, slaty rocks form the lowest or oldest strata, nevertheless, many deviations exist. In many localities entire systems or series of rocky masses, which we find at other places, are wanting: however, this is after all but a local deficiency, and, therefore, of minor importance. We shall see that water was frequently the cause of the destruction of such systems in some localities, while they were preserved in others.

Classification of Formations.

128. The term *formation* in Geology is applied to any portions of the earth's crust, of lesser or greater thickness, which arose under the same contemporary influence. Formations, which, in consequence of their close proximity, stand in mutual relation to each other, are considered by the geologist as connected groups; the separate layers constituting a formation are called its *members*.

129. The coeval formations of the various Neptunic and Plutonic rocks, cannot be easily ascertained, on account of their different external and internal condition, although a subsequent aqueous formation must correspond with a preceding igneous formation. Greenstone and porphyry which have broken

through granite are certainly of later production than granite, as greywacke and coal that overlie the slaty rocks are newer than those rocks.

It would, perhaps, be most conformable to our purpose to designate the different periods of formation by those Plutonic rocks, which were then produced, and thus to classify the total construction of the earth's crust into the periods of the elevation of granite, of greenstone, of porphyry, and melaphyr, of basalt and of volcanos, and treating, intermediately, of the aqueous formations as they were slowly and successively deposited. In all geological systems, however, the terms have been chosen from the stratified rocks, partly because the latter were first examined scientifically, partly because the Plutonic rocks are not everywhere defined with desirable certainty.

130. In the following Table we meet with peculiar terms, some of which are merely accidental, without particular meaning, while others indicate an essential member of the group, as, for instance, the names—marl, red sandstone (new and old), lias, fossiliferous limestone (muschelkalk), &c.

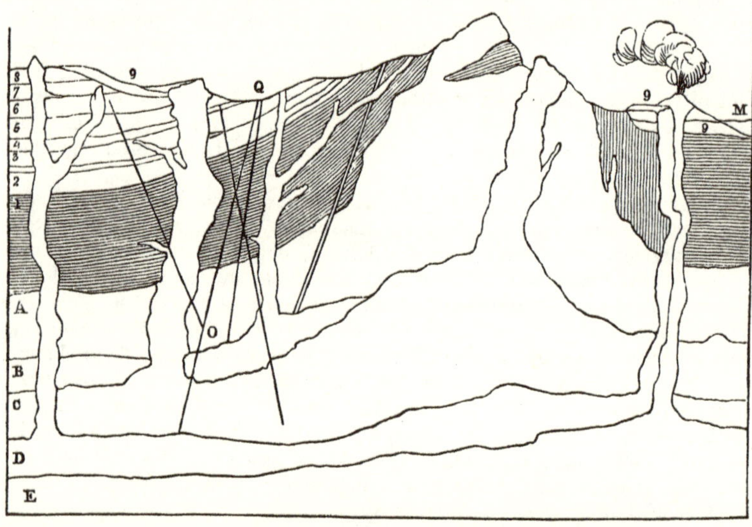

19. Illustration of the Arrangement of the various Groups of Rocks.

Stratified Rocks.		
Quaternary Rocks	9. Diluvial and Alluvial	Deposits.
Tertiary	8. Molasse (comp. § 142).	
Secondary	7. Chalk.	
	6. Jura.	
	5. Trias.	
	4. Zechstein.	
	3. Coal.	
Transition Rocks	2. Greywacke.	
Primitive Rocks	1. Slate.	

Interior mass of the earth.

Massive Rocks.
A. Granite.
B. Greenstone.
C. Porphyry.
D. Basalt.
E Volcanic Rocks.

M. The Sea.
O. Mineral Veins.

Systematic Arrangement of the Formations (beginning with the oldest).

<table>
<tr><th colspan="3">Aqueous Formations
(Neptunic, Normal or Stratified Formation; Stratified Rocks).</th><th colspan="2">Igneous Formations
(Plutonic or Volcanic, Abnormal Formation; Massive Rocks).</th></tr>
<tr><th>Groups.</th><th>Formations.</th><th>Oldest Appellation, according to Werner.</th><th>Groups.</th><th>Most important Rocks of these Groups.</th></tr>
<tr><td>I.
Slate-Group.</td><td>Clay-slate, Mica-slate, Gneiss.</td><td>1.
Primitive Rocks.</td><td>A.
Granite-Group.</td><td>Granite, Granulite, Syenite.</td></tr>
<tr><td>II.
Greywacke-Group.</td><td>Upper and Lower Greywacke.</td><td rowspan="2">2.
Transition Rocks.</td><td rowspan="2">B.
Greenstone-Group.</td><td rowspan="2">Greenstone, Serpentine.</td></tr>
<tr><td>III.
Carboniferous Group.</td><td>Old Red Sandstone, Coal-beds, Mountain-Limestone, Coal Sandstone.</td></tr>
<tr><td>IV.
Zechstein-Group.</td><td>Zechstein.</td><td rowspan="4">3.
Secondary or Stratified Rocks.</td><td rowspan="2">C.
Porphyry-Group.</td><td rowspan="2">Felsite-porphyry, Pitchstone-porphyry, Melaphyr.</td></tr>
<tr><td>V.
Trias-Group.</td><td>Keuper, Muschelkalk, Variegated Sandstone.</td></tr>
<tr><td>VI.
Jura-Group.</td><td>Jura and Lias.</td><td rowspan="2">D.
Basalt-Group.</td><td rowspan="2">Basalt, Phonolite, Trachyte.</td></tr>
<tr><td>VII.
Chalk-Group.</td><td>Chalk, Green Sand, Weald.</td></tr>
<tr><td>VIII.
Molasse-Group.</td><td>Upper Brown Coal, Coarse Limestone, Lower Brown Coal.</td><td>4.
Tertiary Rocks.</td><td rowspan="2">E.
Volcanic Group.</td><td rowspan="2">Lava, Scoria, Volcanic Mud.</td></tr>
<tr><td>IX.
Diluvial and Alluvial Groups.</td><td>Alluvial and Diluvial.</td><td>5.
Quaternary Rocks.</td></tr>
</table>

20. Configuration and Arrangement of the various species of Rocks.

Fig. 20 will afford a general idea of the configuration and arrangement of the various species of rock and veins.

1. Granite.	13. Shale.
2. Gneiss.	14. Calcareous Sandstone.
3. Mica-slate.	15. Ironstone.
4. Syenite.	16. Basalt.
5. Serpentine.	17. Coal.
6. Porphyry.	18. Gypsum.
7. Granular Marble.	19. Rock Salt.
8. Chlorite Slate.	20. Chalk.
9. Quartz Rock.	21. Amygdaloid.
10. Greywacke.	A A. Primary Mountains.
11. Sandstone.	B B. Secondary Mountains.
12. Limestone.	*a a*. Veins.

131. In the study of the stratified rocks the only correct method will be to proceed from the oldest to the most recent formation: first, because this method corresponds with the progress of the development of the earth and of its products; and, secondly, because the description of later conglomerates, if they contain displaced fragments of older stratified rocks, previously undescribed, could not be rendered perfectly clear. By inverting the progress from the more recent to the older groups, the former would appear as if suspended in the air, because the supporting strata, on which they are resting, would not be known.

A.—AQUEOUS FORMATIONS.

Neptunic, Normal, or Stratified Formations.

1st Group—SLATES.

Primitive Rocks.

132. The slate-group has been entered in the table of classification § 130 amongst the aqueous formations, although, from the way in which it originated, we ought perhaps to class it amongst the igneous formations. In that case it ought to precede the granite-group. We class the slates among the stratified rocks because they were designated in § 115 as the *first compact layer* or crust of the once entirely fluid globe, which was however soon broken through by granite. Hence the slate rocks ought to be met with everywhere, if immense bodies of the stratified formations had not covered them in. They are however distributed over the whole surface of the earth and constitute the principal part of a great number of mountains.

Other massive rocks frequently penetrate through this slate-group, especially greenstone, porphyry, and granite. They contain, not unfrequently, veins of ore.

The three principal kinds of this group are clay-slate, mica-slate, and gneiss.

Clay-slate (§ 84), of which the common roofing slate is the purest form, occurs in great variety. It is not so rich in mineral veins, and is less generally distributed than the other two kinds. Great masses of it occur in Wales, in Cumberland, about Loch Lomond, in Scotland, and in many parts of Germany.

133. Mica-slate (§ 85) is very important from the mighty masses which

exist of it. This rock forms large mountains with projecting ridges or craggy tops and precipitous ravines. A great part of the *Swiss* and *Tyrol* Alps

21. Mountains of Mica-slate and Granite, Glen Sannox, Island of Arran.

consists of this rock, which is moreover the prevailing constituent of the *Sudets*, the *Riesen-*, *Erze-*, and *Fichtel-Gebirge.* It is very abundant in Scotland in the mountains that extend from Argyllshire up towards Aberdeen. In the neighbourhood of the places where granite and porphyry break through it, it contains rich veins of ore, which lead to important mining operations.

22. Contorted Mica-slate, Island of Arran.

Gneiss, which holds an intermediate position between mica-slate and granite, occurs in a great variety of forms, and is rich in mineral veins, particularly where it is penetrated by porphyry. This rock forms entire mountains in many parts of the continent of Europe, especially in the alpine districts, where it is commonly associated with granite. It is also abundant in Scotland.

2ND GROUP—GREYWACKE.

Transition Rocks.

134. The term transition rocks, applied to this group, shows, that we have arrived at the confines of the decidedly-stratified formations. These rocks exhibit, in fact, the character of the first crust of the earth, which we have, in § 118, designated as primitive rocks.

The most important members of this group are *greywacke-slate* and *greywacke-sandstone*, which are associated, particularly in the upper parts, with masses of limestone and dolomite. The name of this group is derived from a species of finely-grained sandstone, of grey colour, detached and compact pieces of which lying about on the fields are called " Wacken."

Greywacke has been found distributed in large masses through various parts of Europe, especially in the interior of Bohemia and in the Tyrolese Alps. It occurs likewise in several other quarters of the globe. It is abundant in the south of Scotland at Leadhills. The valleys of the greywacke-group are mostly very winding, as, for instance, that of the Mosel and of the Aar.

The *greywacke-slates* constitute a part of the slaty mountains of the Rhine, and form, in some places, a transition into common slate, fit for roofing. This formation contains, especially in England, *Anthracite* (§ 30), a coal that takes fire with difficulty, and possesses a perfect mineral appearance.

Fossils are found abundantly in the upper members of this group, while the lower contain but few. They are chiefly Polypi, Mollusca, and the so-called *Trilobites*, *i. e.*, extinct crustacea of the isapodous and decapodous families. Fishes and plants appear here more rarely.

3RD GROUP—COAL FORMATION.

135. We have now to consider one of the most important of the various formations, namely, that which includes coal as its essential member, a mineral which as fuel has become indispensable to man for domestic and industrial purposes. This group begins with a coarse conglomerate, consisting of the fragments of older rocks, and never containing basalt, limestone, or flints, and which, on account of its peculiar colour, is called *Old Red Sandstone* (Rothliegendes). This attains to the thickness of even 3,000 feet, and occurs sometimes on the flanks of high mountains, and sometimes constituting by itself mountainous masses, as in the Thuringian forest and the Hartz mountains. Very few impressions of plants are found in this rock.

Upon this old red sandstone follows the *coal formation* properly speaking. It consists of beds of coal from a few inches to 20 feet (and rarely more than 40 feet) in thickness, and between which there is frequently interposed a peculiar *sandstone* (coal sandstone, new red sandstone), and a dark coloured *slate-clay* (coal schiefer). In this order from 8 to 120 beds of coal frequently overlie each other, of which however only the few thicker ones repay the trouble of working. Beneath the coal lies the *greywacke* of the preceding group. The sandstone is exceedingly diversified, and often of fine quality, and well adapted for building purposes.

The outcropping of the coal formation at the surface of the earth seems, in some measure, to depend upon the rising of the mountains; for, in extensive plains, it is generally not discovered, or, in such localities, the beds are too far below the surface to be discoverable or even to be reached by boring.

Coal seems likewise not to have been formed equally in all places during the period in which it originated. The remains of plants found in this stratum lead to the inference, that, during that period there existed an exceedingly vigorous and crowded vegetation, principally confined to equisetaceæ and ferns, of which the *Sphenopteris Hœninghaussii* (fig. 23), *Pecopteris aquilina* (fig. 24), and *Neuropteris Loshii* (fig. 25), are amongst the most beautiful that have yet

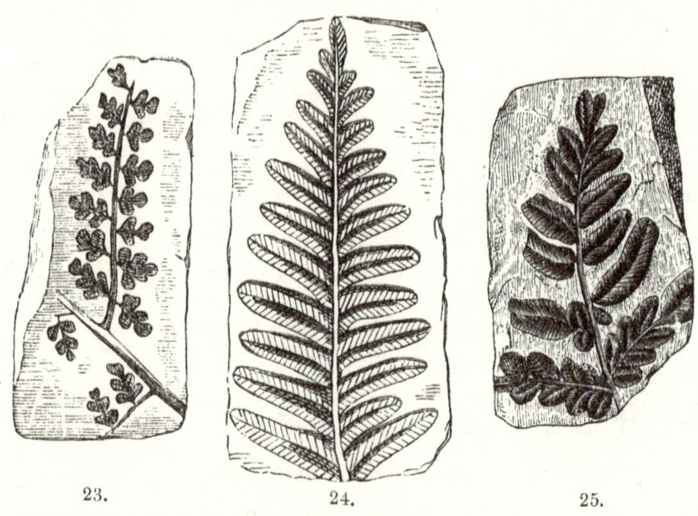

23. 24. 25.

been discovered. These remarkable plants must have presented an aspect essentially different from that of our present forests. Most likely, however, such coverings of vegetation were not everywhere equally crowded together to give rise by their destruction to beds of coal. Hence it is very possible, nay, even probable, that, in some localities, the other members of this group may be found in succession without any beds of coal between them.

It has been generally observed, that the beds of coal are partially surrounded by hills as in a sort of trough, similar to the position of the molasse (§ 142), from which it would appear, as if those forests of vegetation were particularly developed only within these mountain gorges, and could only have formed considerable deposits of coal at those places. The preceding statements afford indications to guide us in searching for coal wherever we may suspect its presence. If the part of the country be composed of primitive mountains or of Plutonic rocks, which we have designated in fig 19 under letters A to E, we may, with certainty, conclude on the absence of coal. If stratified formations of great thickness be present, it is not likely that we shall find coal at an accessible depth; such is more probable, however, where the aqueous formations, bordering upon the massive rocks, have been elevated and uplifted by these, in such a manner that the lower strata approach the surface or even become exposed to our view.

Our search after coal is particularly encouraged by the presence of red sandstone and greywacke, because these formations generally border on beds of coal. If, moreover, the surrounding hills of massive rock should form a basin, the hope of finding coal is all the more well founded, and repeated boring should be resorted to.

136. The coal-beds of Germany are not very numerous, nor is the coal of the best quality. It is in Great Britain that the best coal is most abundantly found, particularly in the neighbourhood of Newcastle-on-Tyne, in Staffordshire, and in Lanarkshire. Coal is also met with in Belgium and the neighbouring parts of France, and at Dombrowa, in Poland. The members of the coal-group have been found generally spread over America, Asia, and even Australia. In South America coal was discovered by Humboldt at a height of 8,000 feet above the level of the sea. The total amount of coal raised annually in Europe exceeds 700 millions of cwts., of which England alone contributes about 450, and Germany about 40 millions.

The *mountain limestone*, which accompanies the coal formation, is a mineral of considerable importance. It usually includes extensive metallic deposits. This is the case in Belgium, in Derbyshire, and in Scotland. Near Glasgow, it is accompanied, not only by coal, but by immense masses of clay ironstone, and in the smelting of iron from that ore, the carboniferous limestone is used as a flux. It forms very beautiful dark-coloured marbles which admit of a fine polish. Organic remains are very abundant in it, of which the following may be cited as characteristic specimens :—

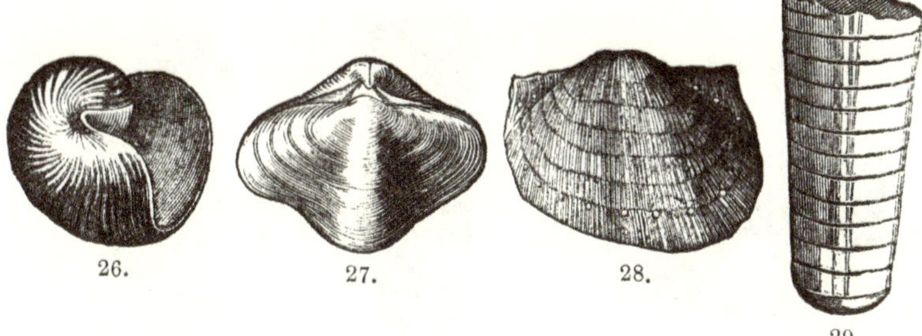

26. Bellerophon costatus.
27. Spirifer glaber.
28. Productus Martini.
29. Orthoceras lateralis.

4TH GROUP—ZECHSTEIN.

137. Of all the formations constituting the crust of the earth, that of Zechstein has been found up to the present time the least distributed. In the north-east of Germany, and especially in the county of Mansfeld, in Saxony, between the sandstone of the preceding group and the conglomerate of the following, there lies, sharply separated, this group, the most essential member of which is a dark bituminous marl-slate rich in copper ores, whence it has

obtained the name of *Copper-Slate.* This is worked in many mines. The Zechstein-group contains but few species of fossils; but these few occur in great quantities: they are all inhabitants of the sea, as corals, shells, and fish.

The upper members of the Zechstein formation occasionally contain *gypsum,* which occurs in some localities in considerable masses, as, for instance, on the south of the Hartz mountains. It is often also accompanied by *rock-salt,* the salt works of Northern Germany depend, therefore, upon the produce of the Zechstein formation. In the neighbourhood of Eisleben and Eisenach many caverns are found within the gypsum beds, which probably arose from the previous existence of rock-salt in them, which, in course of time, was removed by the action of water.

5TH GROUP—TRIAS.

138. The name of this group is derived from its consisting of three members. They occur in Thuringia and Swabia; the entire Black Forest belongs to it, as do the opposite Vosges Mountains, between which the Rhine has excavated its gigantic bed.

Gypsum and *Rock-salt* are characteristic of this group, the upper member of which, called *Keuper,* contains them in great abundance. Thus in Wirtemberg all the salt-works are supplied from this member, as likewise are those at Halle, Fredericshall, Dürrheim, Wimpfen, &c.

Another member of this group is *Muschelkalk,* or shell-limestone, thus called from the great quantities of the shells it contains in separate layers.

The lowest and most abundant member of this group is the *Variegated Sandstone,* which is of a red, yellow, and white colour, and is very abundantly distributed over the continent of Europe. The thickness of its beds varies from 400 to 600 feet, and occasionally reaches to 1,000 feet.

The decrease of fossils in the trias-group is very remarkable. The keuper

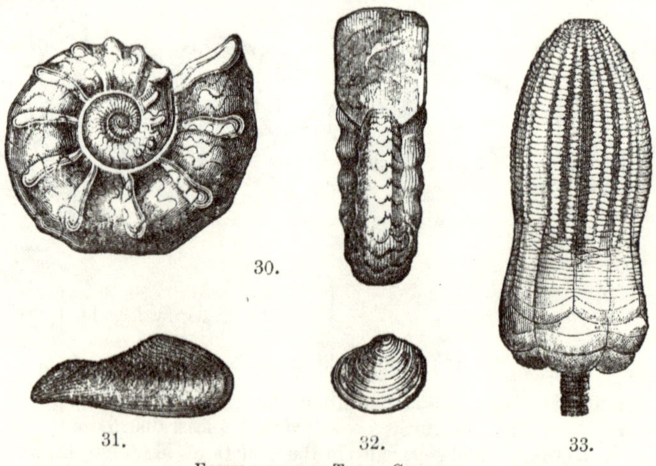

FOSSILS OF THE TRIAS-GROUP.

30. Ammonites nodosus.
31. Avicula socialis.
32. Possidonia minuta.
33. Encrinites moniliformis.

and the variegated sandstone are especially poor in these remains. They are numerous in fossiliferous lime (muschelkalk), but less rich in the number of species than in Jura. Bivalve shells predominate, and the ammonites and belemnites, so frequent in the latter, are here entirely wanting. We must mention, however, the ceratites, as belonging exclusively to muschelkalk. Of fossil plants we find ferns, and the various species of equisetaceæ, in all the members of this group, down to the variegated sandstone. Remains of fish and amphibious animals are rarely met with.

In certain layers of the variegated sandstone there have been discovered hardened footprints, of which it is doubtful whether they belong to mammalia, to birds, or to amphibious animals, which latter is the most probable. At Hessberg, near Hildberghausen, in Saxony, unmistakable impressions of the feet of quadrupeds have been discovered in the grey sandstone of that neighbourhood. (See fig. 34.)

34.

6th Group—JURA.

139. The Jura mountains, which rise from 4,000 to 5,000 feet, have given their name to this formation, which has been found pretty abundantly distributed all over Europe. Limestone is its principal member, occurring alternately with dolomite, marl, clay, and sandstone. In the upper strata a lighter coloured limestone prevails, which turns white when exposed to the air, and contains corals, while among the lower strata dark-coloured limestones and marls are predominant.

In Germany, the Swabian Alps, extending through Bavaria and Franconia up to Saxony, belong especially to the Jura-group. This formation is famous for its numerous caverns, containing deposits of bones. It also contains those compact tabular limestones which are important in their application as lithographic stones. These are found at Solenhofen in Pappenheim.

35. Ammonites Bucklandi.

36. Belemnites mucronatus.

The lowest member of the Jura-formation has received its name, *lias*, from the corruption of the word layers.

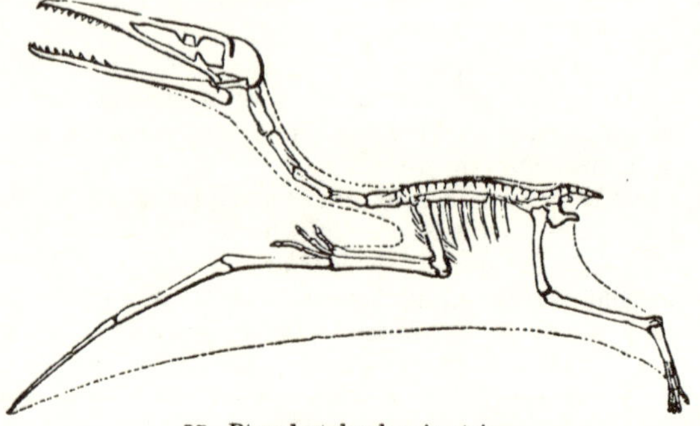

37. Pterodactylus longirostris.

Fossils are exceedingly plentiful throughout the entire formation, particularly mollusca, amongst which there are many *Ammonites* (fig. 35) and *Belemnites* (fig. 36), fish and saurians; of which the winged lizard (*Pterodactylus*, fig. 37) is perhaps the most remarkable member. In the lower strata marine plants are found.

The soils produced from the Jura-formation are fruitful, with the exception of those from the limestone and dolomite mountains.

7th Group—CHALK.

140. While the formations of the preceding groups make their appearance more in dispersed localities, and particularly where the natural conditions of alluvial and diluvial action existed on a more or less grand scale, we find the members of the cretaceous group much more independently and continuously spread. It consists of a defined series of Lime-, Marl-, Sand-, and Clay-Strata, the uppermost of which contain the fossils of marine animals, and the lowermost those of terrestrial plants and of fresh-water animals.

Chalk includes a series of groups, to which the Zechstein, Trias, and Jura groups belong, and which Werner designated as *Secondary* mountain formation. The most striking characteristic of the secondary mountain formation is the absence of the fossils of birds and mammalia, which indicates that it originated under physical conditions essentially different from the later and present formations.

The Chalk-group has been found not only in almost all countries of Europe, but also in various parts of Asia, Africa, and America. Europe, during its formation, seems to have been almost entirely covered by the sea. The rocks of this group form a hilly, or undulated country, without producing high mountains.

141. The most distinguished and characteristic member of the group is *Chalk*, which attains a thickness of from 600 to 900 feet. It passes from white chalk into chalk-marl and limestone, of different degrees of hardness and impurity. Chalk-soil is generally barren, and particularly in France, where there are extensive plains of high and waste table-land entirely covered with it.

It is remarkable that *Flints* are invariably associated with chalk, which encloses them as nodules of various sizes and forms. A closer investigation proves them to be agglomerates of the silicious shells of infusoria. The fossils of this group are exceedingly numerous, especially those of deep-sea animals.

Among the lower members of the chalk-group important strata of sandstone make their appearance, known in England as *greensand*, from being coloured by grains of green earth. In Germany it is termed Quadersandstein, on account of its splitting naturally into large square masses. It is found on the surface of the earth, particularly in Saxony, where it forms the remarkable and picturesque ravines and precipitous rocks of the district near Dresden, known as the Saxon Switzerland.

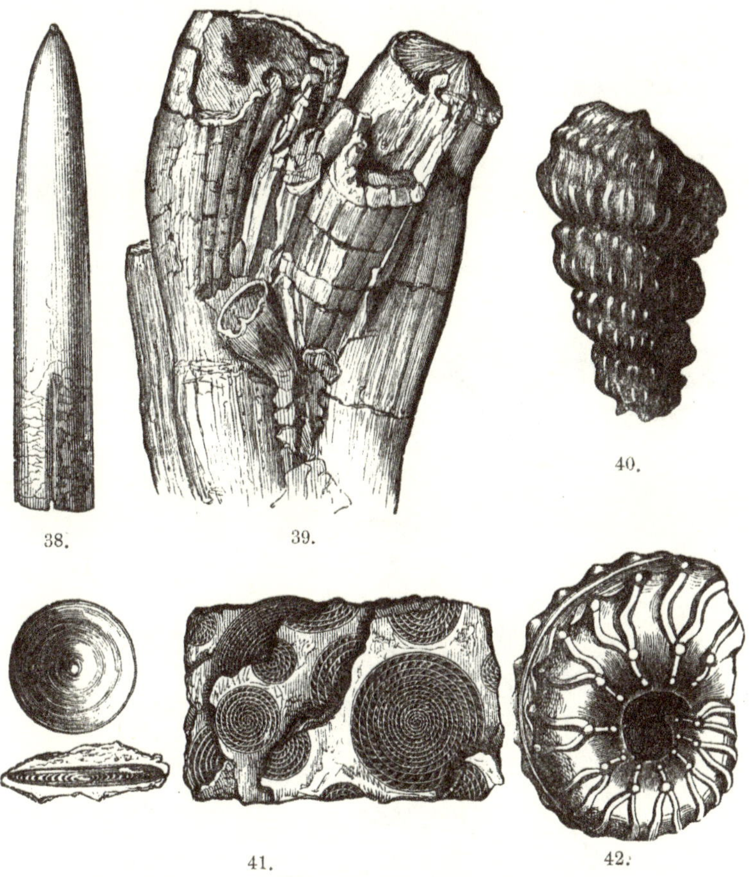

FOSSILS OF THE CHALK-GROUP.

38. Belemnites mucronatus.
39. Hippurites organisans.
40. Turrilites costata.
41. Nummulitic Limestone of the Pyrenees.
42. Ammonites varians.

8th Group—MOLASSE.

Tertiary Rocks.

142. The appellation of this group is derived from a coarse and loose sandstone belonging to it, and occurring in Switzerland, where it is called Molasse: it contains frequently large fragments of other rocks, firmly cemented to a compact mass, named "*Nagelfluh*," and which rises, for instance, on the Rigi, to a height of 6,000 feet. Alternating with strata of *Brown-Coal* and *Calcareous Rocks*, this group forms the margin of the Alps.

During the period when this group was formed, several large bays or gulfs of the sea seem to have been gradually filled up, in which sand, gravel, and marl, with fossils of fresh-water animals, form the principal portion of the upper strata, while in the middle strata a coarsely-grained limestone, termed *Grobkalk*, and an admixture of granular green earth, and a variety of terrestrial and marine fossils prevail. The lower strata are composed of *Clay* and *Brown-coal*. In different localities, however, various deviations occur from this succession.

It is worthy of remark, that several capitals, such as London, Vienna, Mayence, and Paris, are situated in the centre of such filled-up basins. Amongst the fossils of the Mayence basin, the Dinotherium is the most remarkable which has yet been discovered. It is a gigantic animal, similar to an elephant, with two large tusks curved downwards from the lower jaw (fig. 43). In the London basin clay predominates, whilst the basin of Paris furnishes an

43. Dinotherium giganteum.

44. Fauna of the Epoch of the Paris Basin.

a Palæotherium magnum. | *c* Anoplotherium commune.
b Palæotherium minus. | *d* Crocodile.

excellent material for millstones. In the gypsum quarries, near Paris, a large number of marine fossils occur, amongst which about 1,400 extinct species of mollusca have been enumerated.

143. Except in Switzerland, the Molasse does not rise to any considerable altitude. In the north of Germany, in Bohemia, in Wetteravia, &c., the Brown-coal formations predominate, while the middle stratum of Grobkalk is wanting. In the place of the latter a characteristic associate of the lower section deserves our notice; it is a *Sandstone*, distinguished by its great compactness, and as being distributed in large blocks, often strikingly rounded off, over all the north of Germany.

Brown-coal occurs more frequently in level countries and nearer the surface than in the higher grounds, where greater masses of alluvial and diluvial deposits cover it, though even there it is found sometimes uplifted to the surface by massive rocks. In the neighbourhood of Basalts it is considerably altered, probably from the influence of heat. Its ligneous structure disappears almost entirely, and it assumes greater resemblance to ordinary coal. (Chem., § 166.)

It has already been mentioned that well-preserved trunks of trees, leaves, fruits, and also amber enclosing insects, &c., are found in the beds of Brown-coal. Earthy Brown-coal, containing an admixture of clay and sulphide of iron, is worked for alum. (Chem., § 86.)

9TH GROUP—ALLUVIAL AND DILUVIAL DEPOSITS.

144. ALLUVIAL FORMATION, or a deposition of soil from water, still takes place every day under our own observation. Brooks and rivers continually tear away from mountains and valleys more or less of their marginal projections, in proportion to the solidity of the rocks and the power of the fall of water. Thus the elevations of the earth are continually, though imperceptibly, diminished.

The dislodged particles are deposited again in the state of mud, gravel, and pebbles, wherever the streams flow more calmly. Amongst these we find such mineral substances as were distributed in veins through the mountains, and which, on account of their greater specific gravity, were deposited sooner than other less ponderous minerals. Thus gold and precious stones, and also tin-ore, are congregated in many localities of the alluvial and diluvial formations, and may there be searched for with success, while in the mountains whence they come they would be far more difficult to find.

The greatest alluvial deposits from the mud and sand of great rivers are the so-called *Deltas*, triangular islands formed at the mouths of those rivers, and dividing them into many branches, as is the case with the Nile, the Rhine, and the Danube.

Great lakes have been gradually filled up with alluvial deposits.

The sea also continually destroys one line of coast and reconstructs another, and in some localities the formation of a new marine sandstone or limestone has been observed going on gradually from the deposits of evaporated sea-water, and from the remains of finely divided shells. This is the only kind of rock which has hitherto been found to contain the remains of man; a human skeleton having been discovered embedded in this rock on the island of Guadaloupe.

Calcareous Tufa, a rock of by no means inconsiderable extent, belongs likewise to the present period. The carbonate of lime copiously held in solution

45. Action of the Waves on Precipitous Rocks.

46. 47.

Examples of Rocks worn by the Sea.

by the waters of many brooks, lakes, and swamps, containing an excess of carbonic acid, is deposited, when a portion of the carbonic acid escapes into the air (Chemistry, § 52). A coating of carbonate of lime is thus deposited upon all the objects in the water, a rock being gradually formed, which is at first loose and soft, but hardens by exposure to the air, and in this state forms an excellent building material. Such is the famous *Travertine* found in a swamp near San Philippo, in the neighbourhood of Rome, where a layer of this rock attaining a depth of 30 feet has been formed within 20 years. Silicious springs, like those near Karlsbad, and the famous hot-springs of Iceland, the Geysers (fig. 48), deposit *Silicious Sinter*. Moreover, the layers of *Bog Iron-ore* (*Raseneisenerz*) deposited from chalybeate waters, and the saline crusts formed on the shores of the sea, or on the banks of lakes, marshes, and swamps, by their partially drying up, are by no means inconsiderable.

145. Of greater importance, however, are the *Turf* or *Peat Bogs*, the origin of which, falling within historical times, has already been described in the Chemical section of this work (§ 165). They occupy the lower levels, such as the plains of Ireland, Holland, Prussia, Hanover, and Denmark. Sometimes considerable patches are found in hollows on the summits of primitive mountains. Weapons, and other articles made by man, are sometimes found deeply imbedded in peat bogs,—for example, Celtic weapons; and on one occasion,

the wooden bridge, constructed by Germanicus when penetrating through the Netherlands into Germany, was so found. The origin of Peat may be traced

48. The Geysers of Iceland.

back to the older period of diluvial and molasse formations, forming there a transition into Brown-coal.

The *Beds of Infusoria* must be considered in a similar light. An invisible world of the most minute animalculæ, with shells or shields around them, consisting of silicic acid, with the remains of an innumerable host of Infusoria, are deposited in layers, which form a friable mass of silicious rocks, known by the names of *Tripoli*, *Polishing Slate*, and *Kieselguhr*. The annexed diagram (fig. 49) represents a few of the best-defined species which have been recognised by M. Ehrenberg, who has calculated that the space of a single cubic inch would contain upwards of 35,000 millions of such remains. In the ocean this formation is represented by the beds of corals which are built up from the bottom by Polypi, and gradually approach the surface of the water with their calcareous

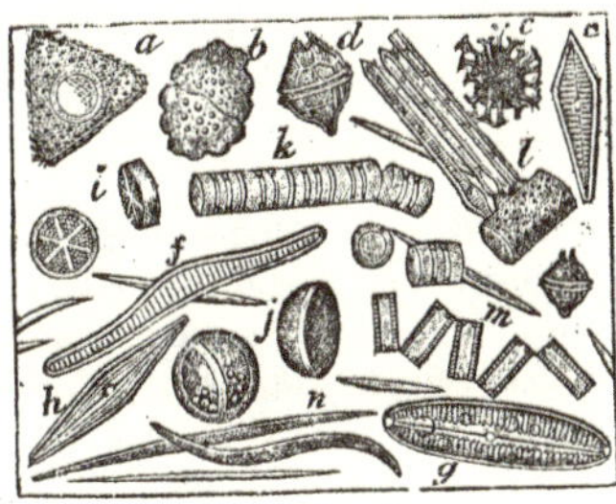

49. Infusoria.

a Desmidium apiculosum.
b Euastrum verrucosum.
c Xantidium ramosum.
d Peridinium pyrophorum.
e Gomphomena lanceolata.
f Hemanthidium arcus.
g Pinnularia dactylus.
h Navicula viridis.
i Actinocyclus senarius.
j Pixidula prisca.
k Gallionella distans.
l Synedra ulna.
m Bacillaria vulgaris.
n Sponge spiculæ.

ramifications, thus appearing above the surface of the sea as coral reefs, and frequently constituting coral islands (fig. 50), which abound in the Pacific Ocean.

50. Whit-Sunday Island in the Pacific.

Among the Polypi which help to form coral reefs, we may cite the following:—

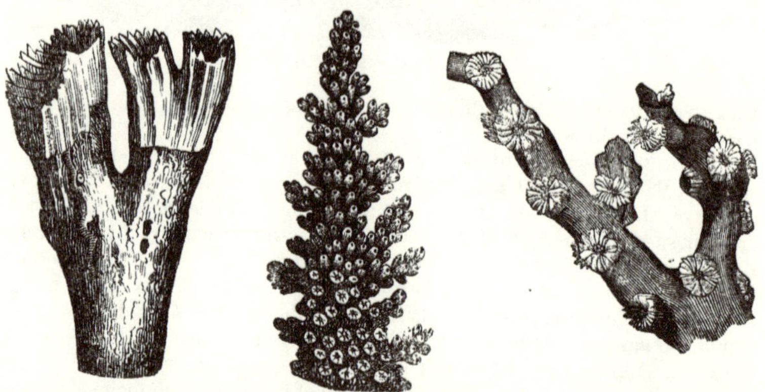

51. Caryophyllea fastigiata. 52. Madrepora muricata. 53. Oculina hirtella.

On the whole, the alluvial formations never reach a considerable thickness above the level of the sea, and they enclose only the remains of still existing plants and animals.

146. Diluvial Formation.—This constitutes even more mighty masses. It arose in pre-historical times as a deposit from a general inundation before the existence of mankind; for it never encloses human skeletons or bones. We find among all nations obscure traditions of mighty floods, which, like the *Deluge*, described in the Bible, had covered a great part of the earth.

The deposits which arose from this flood are of much greater depth than the alluvium deposited from seas and rivers. They are nearly 200 feet thick;

they are generally elevated about 1000 feet, and sometimes as much as 2000 feet above the level of the ocean. The whole of the lower countries of Europe, as well as some plains of smaller extent in its highlands, consist of this formation. Thus the whole valley of the Rhine is filled up with diluvial deposit, which consists of a fruitful marly or sandy loam, which is called *Löss*, because, being too stiff to be washed away gradually by intersecting brooks, it allows itself to be undermined, and breaks downwards vertically, or loosens in masses.

Diluvial deposits enclose many remains of animals, not only of the existing kinds, but also of several extinct species. Among the latter we find particularly large terrestial animals, such as the mammoth, the cavern bear (*Ursus spelæus*), &c. The accumulation of such fossil bones in many caverns is very remarkable; for instance, at Muggendorf in Bavaria, Gailenreuth in Franconia, in the Baumanns and Biels caverns of the *Hartz*, in the Nebel cavern near Tübingen, and in several other localities. These may have arisen from the caverns having been the places of resort of various carnivorous animals, or from the action of the floods carrying the bones thither.

147. Certain migrations of detached masses of rock may likewise have occurred at this period of great floods, as otherwise they would be as inconceivable as they are incongruous with the present state of things. In the great plains of Northern Germany, we find large blocks of rounded stone, principally of granite, lying about singly all over the diluvial deposits, and which we thence call, *Erratic Blocks* or *Boulders*. No granite can be discovered far and wide in their neighbourhood, nor at any depth below the surface. It is certain that these blocks must have been transported over sea, from Scandinavia or Finland, where mountains of the same kind of rock still exist, and it is probably that they were conveyed by immense icebergs, which detached them on breaking up. The descriptions given by northern travellers, of the size of the icebergs still floating about in the polar regions, renders this not at all improbable.

B. IGNEOUS FORMATIONS.

Plutonic and Volcanic: Abnormal Formations. Massive Rocks.

148. In this division we have classed the groups of granite, greenstone, porphyry, basalt, and volcanic rocks, which are indicated in fig. 19, page 334, by the letters A, B, C, D, and E.

The massive rocks, not overlying each other in regular strata, but occurring only wedged, as it were, beside and into each other, it is generally much more difficult to separate the different groups accurately; moreover the fossils which so much facilitate the distinction of the stratified groups, are entirely wanting in these rocks.

The massive rocks distributed over the surface of the earth are more uniform in their constitution than the sedimentary rocks, a circumstance which may be explained by supposing them to have been upheaved from the interior of the earth, and consequently less subjected to external and local influences than the substances of the stratified formations.

A. Granite-Group.

Primitive Rocks.

149. *Granite* was long considered to be the true primitive or fundamental rock, an opinion which extends even beyond the circle of scientific geologists. According to our previous statements, however, we consider it merely as the

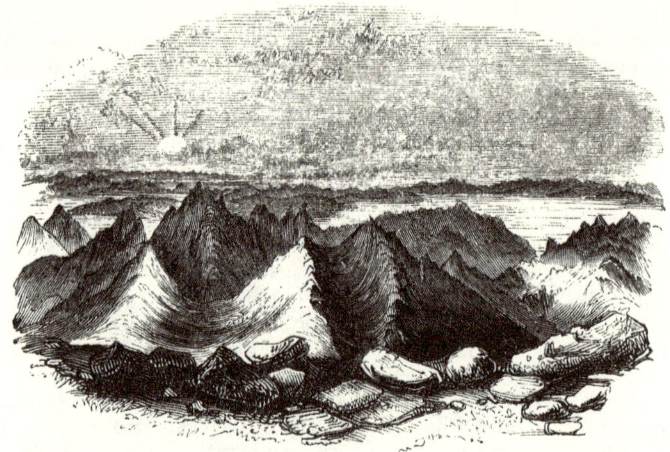

54. Mountains of Granite and Mica-slate, seen from the summit of Goatfell, in the Island of Arran.

first of a series of igneous rocks, which in various subsequent periods, sufficiently remote from each other, broke through the crust of the earth.

55. Granite Boulders, Island of Arran.

This rock occurs likewise in many varieties, of which granite, granulite, and syenite, are geologically considered the more important. *Granite* (§ 87) is less distributed than the slaty rocks. It occurs principally in the form of mountains, and is rarely found in plains. The external configurations of granite are various, but peaked mountains and rugged isolated crags prevail, piled upon each other in great quantities into picturesque groups of apparent ruins. Peculiarly characteristic are the large blocks, like woolsacks, which often abound on the surface of granitic districts. These are large fragments of granite, the angular edges of which, having been worn off by gradual decomposition, they remain as rounded blocks. Mineral veins are not frequent in granite, yet ironstone and tin-ore must be mentioned as occurring in this rock; accidental admixtures of several precious stones and laminæ of gold are likewise occasionally found.

Granite abounds in the north of Scotland, in the island of Arran, in Wales, and in Cornwall.

Granulite (§ 87) is found to a less extent, but under interesting circumstances, at the northern foot of the Erzgebirge, in Germany.

Syenite (§ 88) is found less widely distributed in Europe than granite, but is said to extend over large tracts of country in Chili, and at Mount Sinai.

Syenite is often found ruptured by granite, whence it is thought to be of earlier formation than the latter rock.

56. Corrie-an-Lachan, Island of Arran.

The above cut exhibits another remarkable peculiarity in the scenery of granite mountains, namely the occurrence of a lake that fills what appears to resemble a volcanic crater.*

B. Greenstone-Group.

Trap-Formation.

150. Differing from the rocks of the preceding group, greenstone never occurs in extensive masses, nor forms entire mountains, nor even considerable parts of mountains. It forms, on the contrary, small irregular masses, hillocks, blocks, and intricate veins or dykes, particularly in the substance of granite, slate-rocks, greywacke, and sandstone. In general, greenstone, when it appears on the surface, constitutes small rounded summits, which, in districts of clay-slate, may be recognised even at a distance. The internal cleavage of greenstone is chiefly either nodular or spherical, being rarely seen split in the form of columns and slabs. Of the many varieties of greenstone those of *Diorite* (§ 89) and *Serpentine* (§ 41) occur to the greatest extent. Mineral veins are rarely met with in these rocks, but they contain frequently ores, for instance, of iron, copper, and tin, as accidental components, sometimes in sufficient quantities to render them worthy the attention of the miner.

C. Porphyry-Group.

151. According to Leopold v. Buch, the various porphyries must not merely be considered as a frequent cause of mountainous upheavals, since they also rise up frequently by themselves as considerable mountainous masses. They have been found in all parts of the globe, under the same relative arrangement, breaking in trunk-like masses, or in wide-spreading veins, through the granite and slate formations, and through the greywacke and carboniferous groups of the secondary rocks. In their external appearance, the porphyries appear to be peculiarly adapted for the formation of rocky mountains, and frequently they constitute isolated hills in the midst of other rocks. They cleave into angular fragments, and frequently split into multiform pillars and

* We borrow from Professor Ramsay the following description of such a scene :—

"Before descending to the coast, let the geologist turn aside to see a solitary mountain tarn, in the silent recesses of Beina Mhorroinn. This little sheet of water is by far the most picturesque of all the lochs of Arran, and is situated deep in a hollow, called Corrie-an-Lachan. The place is perfectly lonely ; not a tree is near ; and except the brown heath on its margin, and a few stunted rushes by the brook, the surrounding hills are almost bare of vegetation. The water is dark and deep, and the stormy blasts of the mountain never reach its still and unruffled surface. From its edge, on all sides but that towards the sea, rise the naked hills, whose sides are either formed of massive granite blocks, which, though surely yielding to decay, yet offer a stronger resistance to the destroying influences of time than the softer portions of the mountain, where the decomposing rock may almost be seen slowly crumbling away.

"A remarkable feature of the granite hills of Arran, is the Corries (one of which is represented in the cut). These may be frequently observed in the ridge between Brodick and Sannox, and in the hills of the interior. They generally present the appearance of a volcanic crater, part of one side of which has disappeared ; and the masses of granite which compose the encircling hills are frequently arranged in layers diverging from the centre of the Corrie according to the angle of inclination of the hill. For obvious reasons it will be evident to the most inexperienced observer, that there is no analogy between the Corries (Corrie = Cauldron ?) and modern volcanic craters ; and it is probable that they owe their origin to the softer nature and earlier decay of the rock, with which at remote periods they may even have been nearly filled. May not even the great glens owe their origin to the same cause ?"—*Geology of the Island of Arran*, 8vo, Glasgow, 1841, page 50.

slabs. In their point of contact with other rocks, *Breccias* frequently occur (§ 96).

A great many varieties of porphyry exist, amongst which, *pitchstone porphyry*, *melaphyr*, and *amygdaloid*, are the most important.

Pitchstone porphyry occurs only in isolated masses. *Melaphyr* and *amygdaloid* are more widely distributed; they do not however constitute extensive districts, but form small trunk-like masses and irregular veins in Upper Silesia, Bohemia, Saxony, Scotland, and in many other localities.

D. Basaltic Group.

152. This upheaved group exhibits so decided a character that it is easily recognisable even by the unpractised eye. Being of much later date than most of the secondary formations, or than the above-named massive rocks, it is found to have broken through them and penetrated even up to the molasse-group. Only the diluvial and alluvial formations have been formed since the appearance of the basalt.

Basaltic rocks frequently form lines of spreading hilly country, independent of chains of mountains; or, very characteristically, they constitute single dome-shaped elevations or conical hills in the flat regions of the stratified formations. They are distributed all over the globe, and in Germany they form a very remarkable basaltic zone, running from east to west.

Isolated basaltic cones sometimes attain a height of 1,000 feet, and present to the eye the most varied and graceful cleavage; the basalt itself consisting mostly of regular hexagonal or pentagonal columns.

The more important varieties are *Phonolite* (§ 93) and *Trachyte* (§ 94); which are however rather rare, and occur mostly associated with the common basalt.

57. Basaltic Columns, Island of Arran.

The rocks of this group are not penetrated by veins of ore.

Wherever the basaltic rocks border on other kinds of rocks, the most remarkable phenomena have originated at the period of their upheaval as a glowing liquid mass. In such localities these latter rocks have undergone great alteration still distinctly visible, being partly fused or reduced to mere slag, similar to the effects of volcanos, still in activity, or to the process of our smelting furnaces, where such igneous formations are constantly produced on a smaller scale. The appearance produced by the junction of the trap-rocks with sandstone, slate-rocks, &c., can be easily examined at Dunoon, in the Island of Bute, or in numerous other localities on the Clyde.

58. Sunk Trap Dyke. 59. Raised Trap Dyke.

When a trap dyke is more durable than the penetrated strata, the rock which it traverses being worn away by the action of the elements, the trap dyke is left above its surface in the form of a wall, fig. 59. But when the dyke is more perishable than the rock which it pierces, the trap decomposes and leaves a hollow, bounded on each side by a perpendicular wall of sandstone, fig. 58.

60. Basaltic columns on the border of the river Volant in Ardèche, France.

E. Volcanic Group.

153. We have already explicitly described in § 123 the activity and the influence of volcanos upon their surrounding neighbourhood. According to modern views, all upheaved massive rocks might be considered as extinct volcanos, some of which are of immense extent. However, it is only with the group of basalt, immediately preceding the volcanic group, that we find a considerable approximation in character to the present volcanic formations.

A characteristic feature of volcanos is the conical form of their summits, which appear sometimes isolated and sometimes in groups or chains. A farther characteristic is the formation of a funnel-shaped crater at their summits. The rocks which we meet with in volcanos, and in their immediate neighbourhood, consist of *lava*, *slags*, and *trachyte* (§ 94), in which no mineral veins are present.

61. Extinct Volcanos of Auvergne.

CONCLUSION.

154. On taking a retrospective glance at what has been stated under the heads of Mineralogy and Geology, we find ourselves progressing most remarkably from the minute and elementary to the greatest and most complicated phenomena.

First. Mineralogy teaches us, in the simple mineral specimen, the chemical combinations formed by Nature, the determination of which as well as of their form of crystallisation, is properly considered a part of Chemistry. These minute crystals do not, however, occur merely isolated, but in aggregations of great number united to continuous masses. We also frequently find the crystals of different minerals intermingled and closely united in greater masses, when their definite form of crystallisation is often interfered with by the mechanical actions of friction, pressure, admixture, and by partial or entire fusion or solution. Thus, from the consideration of the simple and compound rocks, Geology leads us on to the contemplation of still greater masses, and their arrangement in successive strata.

155. In describing so many most useful mineral substances, the importance of the science here treated must have become evident to every one.

The mineralogist teaches us not only to *distinguish* such minerals as sulphate of baryta and sulphate of strontia, limestone, salt, sulphur, coal, and the best of ores, so indispensable to man, but he also informs us under what local circumstances we may expect to *find* them.

Besides this, the knowledge of the mineralogist enables him better to judge of the nature of soils produced by decomposition; and, indeed, this knowledge of soils so essential to agriculture has been made the subject of scientific treatment, founded on Mineralogy. Geology, again, has lent its aid for another important purpose,—to procure one of the most indispensable necessaries of life, viz., water. In the section *Physics* (§ 60), it has been shown how this liquid, while endeavouring to find its level, springs up as a fountain, wherever it can force its way. Experience has taught us, however, that we can assist its course in this respect, that we can make channels for it in certain localities, or, in other words, that we may form artificial springs by boring.

ARTESIAN WELLS.

156. The possibility of forming such a well, named *Artesian*, after the department Artois, in France, where the attempt was first made, depends upon certain geological conditions, tolerably well ascertained, according to which a well-informed geologist may easily judge whether in certain localities boring is practicable with the probability of success.

This would be the case under the following circumstances :—

(1.) Water must be continually absorbed and collected on an elevated point, higher than the place where the boring is to be tried.

(2.) This water must, from the nature of the formations below, find access beneath the point of boring by subterraneous channels.

(3.) It must have no other artificial or natural egress, to an amount equal to the quantity at the collecting points, neither at nor below the level of the boring point.

These three general conditions may actually be fulfilled in various ways. Most commonly, however, they are realized in the stratified formations by the peculiar position and alternating quality of the several strata. If, for instance, a sandy stratum, acting as a filter (*a b*, fig. 62,) occupies a somewhat inclined position between two other strata impervious to water, such as clay, the water being absorbed by the superficial parts at *a*, *b*, which may be of very great extent, will penetrate through its whole depth, and finding no egress below on account of the basin-like form of the stratum, as in fig. 62, or from its resting at the lower termination upon a compact rock, the water will collect under sufficient pressure to form an artesian well. The overlying strata need only be bored through, as shown in the centre of the figure, to obtain the desired spring. The passage *f d* in fig. 62, explains the manner in which a natural spring *d* may be supplied with water from a porous bed through a fissure in the rocks. Similar conditions may exist in localities where massive rocks prevail, by means of fissures and cavities, although these are of rarer occurrence, and do not admit of a decided judgment beforehand.

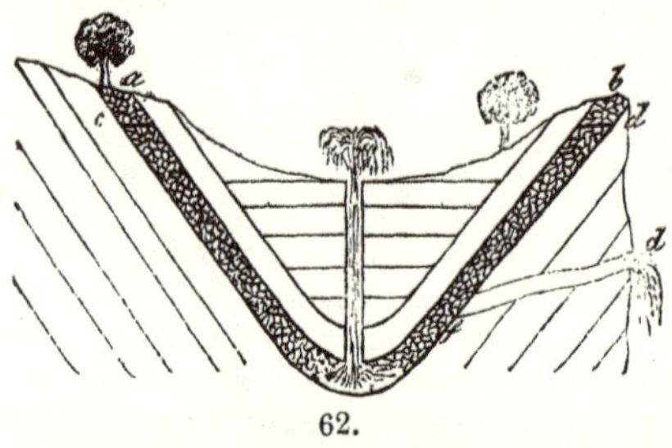

62.

Hence, while in stratified formations, we may predict frequently with great certainty the success of boring for an artesian well, such an undertaking will, on the whole, be very hazardous in localities where slaty or massive rocks predominate, and, consequently, the desired result would not be at all probable.

Artesian wells from a great depth possess a high temperature, as, for instance, the water of the artesian well at Grenelle, near Paris, which is 1,663 Parisian feet (= 540 met.) in depth, possesses a temperature of 28° C. (82·4 F.) This opens up a speculative view of making the immense store of subterraneous heat available for our domestic purposes. Should the stratified formations, from which the artesian spring rises, contain mineral substances soluble in water, in such case it would appear as mineral water. Thus in the Keuper and Zechstein (§ 137 and 138), so rich in beds of rock-salt, saline springs, for the manufacture of common salt, have been frequently found by boring.

MINING.

157. In order to procure for man the comforts and necessaries of life by the assistance of gold and silver, by that of iron, coal, salt, and other minerals, the MINER unceasingly performs his laborious task with steady perseverance.

Miners are generally a poor but an honest and industrious class of people, quiet and earnest at their work, but cheerful and fond of musical entertainment in their hours of recreation. Separate manners, habits, and dress, as well as a peculiar language for everything concerning their occupation, make the miners a characteristic set of men, strongly distinguished from agriculturists, sailors, or townsmen.

With his tools, consisting of a pickaxe, hammer, and crowbar, and provided with a safety lamp, the miner proceeds either to work shafts vertically down into the ground, forming deep pits, or he carries out galleries in horizontal directions, and by combining these two ways he penetrates the rocks in search of ores which run through them in veins or form entire beds in separate strata, as, for instance, coal or rock-salt. Mines are sometimes of immense extent, for some shafts have been sunk to the depth of 3,000 feet: the greatest depth, however, below the level of the sea amounts only to from 1,300 to 1,600 feet, which would make only about $\frac{1}{14800}$ of the radius of our globe (V. Humboldt's Cosmos). The galleries extend in some mines to an astonishing length, as, for instance, the George-gallery in the Harz, which requires three hours to pass through, and the celebrated Christopher-gallery in Salzburg, 10,500 feet long. These galleries, though mostly of a height sufficient for a man just to walk through, frequently admit of access only in a stooping or creeping position.

158. The calling of the miner, besides being very toilsome is, next to that of the sailor, exposed to the greatest amount of danger. There are many mines, in which, out of 1,000 workpeople, an average of 7 annually lose their lives, while about 200 suffer more or less personal injury from accidents. In others, it is stated, that an average of even from 12 to 16, out of 250 people, perish annually. Sometimes, a sudden irruption of water from below or laterally, sometimes the fire-damp (Chem. § 54), which explodes on taking fire, or suffocating gases, especially carbonic acid gas (Chem. § 52), choke damp, prove destructive to them. At times, also, the roof of the mine itself gives way, either from negligence in propping, or from unavoidable concussions, and buries the miners alive. This frequently happens, particularly in South America, where earthquakes are still of common occurrence. All these circumstances contributed much in former times to make the miners a particularly superstitious class of people, abounding in fictions and traditions of jealous mountain sprites, dwarfs, and hobgoblins, dwelling in the interior of the mountains, and watching over their ores and treasures which they grudge mankind, and for taking of which they assail the miner, and seek to do him harm. On the other hand, they believe in benevolent fairies and protecting spirits that aid and assist them.

The progress of science and education has, however, cleared away much of this prejudice and ignorance: the better-informed miner of the present time knows how to distinguish truth from fiction, and while trying to avoid dangers by needful precautions, he puts his trust in God, the ruler of all things.

INDEX.

MINERALOGY.

GEOLOGY.

www.ingramcontent.com/pod-product-compliance
Lightning Source LLC
LaVergne TN
LVHW091638100826
845152LV00005B/88

* 9 7 8 1 4 2 5 5 8 9 0 9 7 *